ONCOLOGY

International Journal of Cancer Research and Treatment

Founded 1948 as 'Oncologia' by H.R. Schinz
Continued by V. Richards (1967–1975), H. Wrba (1976–1992)

KARGER

Printed in Switzerland
on acid-free paper by
Druckerei Schüler AG, Biel

Appears bimonthly:
1 volume per year (6 issues)

ONCOLOGY

Submission: Only original papers written in English are considered and should be sent to:

S. Karger AG
Editorial Dept. 'Oncology'
P.O. Box
CH-4009 Basel (Switzerland)

Manuscripts should be submitted in triplicate (with three sets of illustrations of which one is an original), typewritten double-spaced on one side of the paper, with a wide margin.

Conditions: All manuscripts are subject to editorial review. The editors reserve the right to improve style and, if necessary, return the manuscript for re-writing to the author. Submission of an article for publication implies the transfer of the copyright from the author to the publisher upon acceptance. Accepted papers become the permanent property of 'Oncology' and may not be reproduced by any means, in whole or in part, without the written consent of the publisher. It is the author's responsibility to obtain permission to reproduce illustrations, tables, etc. from other publications.

Arrangement
Title page: The first page of each paper should indicate the title (main title underlined), the authors' names, and the institute where the work was conducted. A short title for use as running head as well as the full address of the author to whom correspondence should be sent are also required.
Full address: The exact postal address complete with postal code must be given at the bottom of the title page. Please also supply Phone and Fax numbers.
Key words: For indexing purposes, a list of 3-10 key words in English is essential.
Abstract: Each paper needs an abstract of up to 10 lines.
Small type: Paragraphs which can or must be set in smaller type (case histories, test methods, etc.) should be indicated with a 'p' (petit) in the margin on the left-hand side.
Footnotes: Avoid footnotes. When essential, they are numbered consecutively and typed at the foot of the appropriate page.
Tables and illustrations: Tables and illustrations (both numbered in Arabic numerals) should be prepared on separate sheets. Tables require a heading and figures a legend, also prepared on a separate sheet. For the reproduction of illustrations, only good drawings and original photographs can be accepted; negatives or photocopies cannot be used. When possible, group several illustrations on one block for reproduction (max. size 181 × 223 mm) or provide crop marks. On the back of each illustration, indicate its number, the author's name, and 'top'. Color illustrations are reproduced at the author's expense.
References: In the text identify references by Arabic numerals [in square brackets]. Material submitted for publication but not yet accepted should be noted as 'unpublished data' and not be included in the reference list. The list of references should include only those publications which are cited in the text. Do not alphabetize; number references in the order in which they are first mentioned in the text. The surnames of the authors followed by initials should be given. There should be no punctuation other than a comma to separate the authors. Cite all authors, 'et al' is not sufficient. Abbreviate journal names according to the Index Medicus system. (Also see International Committee of Medical Journal Editors: Uniform requirements for manuscripts submitted to biomedical journals. Br Med J 1988;296:401-405.)
Examples
(a) Papers published in periodicals: Kauffman HF, van der Heide S, Beaumont F, Blok H, de Vries K: Class-specific antibody determination against *Aspergillus fumigatus* by means of the enzyme-linked immunosorbent assay. III. Comparative study: IgG, IgA, IgM ELISA titers, precipitating antibodies and IgE binding after fractionation of the antigen. Int Arch Allergy Appl Immunol 1986;80:300-306.
(b) Monographs: Matthews DE, Farewell VT: Using and Understanding Medical Statistics. Basel, Karger, 1985.
(c) Edited books: Hardy WD Jr, Essex M: FeLV-induced feline acquired immune deficiency syndrome: A model for human AIDS; in Klein E (ed): Acquired Immunodeficiency Syndrome. Prog Allergy. Basel, Karger, 1986, vol 37, pp 353-376.
Further details: Authors may consult our leaflet 'Rules for the Preparation of Manuscripts', available on request from the Publisher, S. Karger AG, P.O. Box, CH-4009 Basel (Switzerland).

Page charges: There is no page charge for papers of 3 or fewer printed pages (including tables, illustrations and references). Each additional complete or partial page is charged to the author at SFr. 225.– / US$ 150.00. The allotted size of a paper is equal to approx. 9 manuscript pages (including tables, illustrations and references).

Galley proofs: Unless indicated otherwise, galley proofs are sent to the first-named author and should be returned with the least possible delay. Alterations made in galley proofs, other than the correction of printer's errors, are charged to the author. No page proofs are supplied.

Reprints: Order forms and a price list are sent with the galley proofs. Orders submitted after the issue is printed are subject to considerably higher prices. Allow five weeks from date of publication for delivery of reprints.

KARGER

ONCOLOGY

Publication data: 'Oncology' is published 6 times annually. Volume 50 with 6 issues appears in 1993.

Subscription rates: Subscriptions run for a full calendar year. Prices are given per volume.

Personal subscription:

Subscription + postage and handling
SFr. 555.10 + SFr. 16.80 Europe
SFr. 555.10 + SFr. 25.20 Overseas
US$ 370.30 + US$ 16.80
(Must be in the name of, billed to, and paid by an individual. Order must be marked 'personal subscription'.)

Institutional subscription:

Subscription + postage and handling
SFr. 793.– + SFr. 24.– Europe
SFr. 793.– + SFr. 36.– Overseas
US$ 529.00 + US$ 24.00

Airmail postage: SFr. 36.–, US$ 30.00. (Extra per volume.)

Microform edition: SFr. 396.50, US$ 264.50. (Available exclusively to subscribers of the paper edition.)

Advertising: Correspondence and rate requests should be addressed to the Publisher.

Subscription orders: Orders can be placed at agencies, bookstores, directly with the Publisher S. Karger AG, P.O. Box, CH – 4009 Basel (Switzerland) or with any of the following distributors:

Deutschland: S. Karger GmbH, Postfach D – 79095 Freiburg

France: Librairie Luginbühl 36, bd de Latour-Maubourg, F–75007 Paris

Great Britain, Ireland: S. Karger Publishers 11 Ravenscroft Road Beckenham, Kent BR3 4TP

USA: S. Karger Publishers, Inc. 26 West Avon Road, P.O. Box 529 Farmington, CT 06085

India, Bangladesh, Sri Lanka: S. Karger India Subscription Agency B – 5/132 Safdarjung Enclave New Delhi-110029

Japan: Katakura Libri, Inc. 36-9 Hongo 3-chome, Bunkyo-ku, Tokyo 113

Change of address: Both old and new address should be sent to the subscription source.

Single issues and back volumes: Inforr… …an be obtained through the Publisher.

Bibliographic indices: This journal is regularly listed in bibliographic services, including *Current Contents®*.

Book Reviews: Books are accepted for review by special agreement.

F 93378

Clinical Hyperthermia

Selected Papers of the
XVth International Symposium,
Lyon, May 19–22, 1992

Guest Editors
F.N. Gilly,
G. Braillon,
A.C. Sayag,
P.Y. Carry,
A. Brachet,
Lyon

39 figures and 31 tables, 1993

KARGER

S. Karger
Medical and Scientific Publishers
Basel · Freiburg · Paris · London
New York · New Delhi · Bangkok
Singapore · Tokyo · Sydney

Drug Dosage
The authors and the publisher have exerted every effort to ensure that drug selection and dosage set forth in this text are in accord with current recommendations and practice at the time of publication. However, in view of ongoing research, changes in government regulations, and the constant flow of information relating to drug therapy and drug reactions, the reader is urged to check the package insert for each drug for any change in indications and dosage and for added warnings and precautions. This is particularly important when the recommended agent is a new and/or infrequently employed drug.

Printed in Switzerland on acid-free paper by
Graphische Anstalt Schüler AG, Biel
ISBN 3–8055–5854–6

Contents

Oncology 1993;50:333–337

A.C. Sayag[a]
F.N. Gilly[b]
P.Y. Carry[a]
J.P. Perdrix[a]
G. Panteix[c]
A. Brachet[b]
V. Banssillon[a]
G. Braillon[b]

Departments of
[a] Anesthesia,
[b] Surgery, and
[c] Biology, Labo MS 44, Centre Hospitalier Lyon-Sud, Lyon Pierre-Bénite, France

Intraoperative Chemohyperthermia in the Management of Digestive Cancers

A General Review of Literature

Key Words
Chemotherapy
Hyperthermia
Carcinomatosis

Abstract
Intraoperative chemohyperthermia is a new method in the treatment of peritoneal seedings from digestive cancers, which combines surgery, intraperitoneal chemotherapy (mitomycin C and/or cisplatyl) and peritoneal hyperthermia. After a brief reminder on the general principles concerning high temperature action, a review of literature is made: 5 teams have performed this technique. We differentiate the indications, design features and results of each team. The results show a mean survival after 2 years of 35% (in peritoneal carcinomatosis) up to 78% (in gastric serosal invasion, peritoneal seeding free). The best result of the method is the drying up of cancerous ascites, allowing a more comfortable survival.

Introduction

The most discouraging consequence of digestive cancers remains the progressive development of peritoneal seedings against which there is no clearly effective treatment [1–3]. Surgery alone is useless, and general chemotherapy is ineffective. Numerous complementary methods have been evaluated, such as intraperitoneal chemotherapy, radiotherapy and local immunotherapy, without any well-established results. The efficacy of heat against tumors and malignant cells has been well known for years, and has been the subject of numerous studies trying to explain its mechanisms [4–8]. The combination of hyperthermia and chemotherapy was proven to induce an enhancement in the respective anticancerous actions [9–11], so the idea came about to perform triple treatment combining surgery (to reduce the tumoral volume as much as possible), intraperitoneal chemotherapy and local hyperthermia. This treatment was first described and realized by Spratt et al. [12–14] in a patient suffering from a pseudomyxoma peritonei from distal pancreas.

The purpose of this paper is to bring to mind the scientific basis of hyperthermia, to make a synthesis of the different techniques of intraoperative chemohyperthermia (IPCH) and to analyze primary results.

General Principles of IPCH

In vivo and in vitro studies have shown that cancer tissues are more heat sensitive than normal tissue because of their low pH, hypoxic state and abnormal metabolic

A.C. Sayag
Department of Anesthesia
Centre Hospitalier Lyon-Sud
F–69310 Lyon Pierre-Bénite (France)

0030–2414/93/0505–0333
$2.75/0

process. Hyperthermia is an antitumoral agent as soon as it reaches 42.5 °C, acting as a cytotoxic agent for tumoral cells, preferentially during the S phase [4, 5]. It affects membrane stability by denaturing proteins and so induces instability of the phospholipid bilayer. By this means it alters membrane permeability for cations with an increased K^+ efflux, and an enhanced Ca^{2+} and H^+ influx [6, 7]. It inhibits DNA replication, RNA and protein synthesis.

Hyperthermia dysregulates the metabolic process of cell by inhibiting the production of energy and leading to ATP depletion [5].

If we consider the perfusion rate of the whole tumor after a transient improvement in mild hyperthermia, an important decrease is observed as temperature increases leading to a global hypoperfusion ('steal phenomenon') of the tumor inducing edema and necrosis. This low intratumoral blood flow, inducing less heat dissipation and higher intratumoral temperatures, can be reached [4, 9].

The addition of hyperthermia to an anticancerous drug leads to a synergistic enhancement of each other's action and tolerance [10, 11]. An elevated temperature of up to 40–42 °C sensitizes tumoral cells to anticancerous chemotherapy (particularly to alkylating agents). Furthermore hyperthermia induces changes in the pharmacokinetics of the different anticancerous drugs and in their toxicity [10, 13].

The direct introduction in the peritoneal cavity of heat plus anticancerous drug by means of a liquid support (peritoneal dialysis solution) seemed judicious: first it acts directly on the tumor with high extra- and intracellular drug concentrations; second it washes out intraperitoneal free cancer cells, and finally it induces an important portal absorption leading to a more effective action in the vascular portal system [3, 14].

Literature Review

Since the first experience of Spratt et al. [14], 5 teams have performed IPCH (table 1).

Design Features

IPCH was performed under general anesthesia through a median laparotomy. Inflow and outflow drainages were set up inside the peritoneal cavity and connected to the circuit. Intraperitoneal hyperthermia was induced by a heated fluid propelled in the circuit by a roller pump with a 200-ml/min flow rate. Chemotherapy was added to the liquid. Inflow, outflow and intraperitoneal thermic probes were connected to a thermic reader to manage heat transfers during IPCH.

Table 1. Literature review

Reference	Material	Stage
Koga et al. [17]	59 gastric cancers	T_3M_0
Kaibara et al. [18]	42 gastric cancers	T_3M_0
Fujimoto et al. [19]	15 gastric cancers	T_3M_{p+}
Fujimoto et al. [20]	41 gastric cancers	T_3M_{p+}
Fujimoto et al. [21]	31 gastric cancers	T_3M_{p+}
Fujimura et al. [22]	41 gastric cancers	T_3M_{p+}
Yonemura et al. [23]		
Gilly et al. [15, 16]	21 digestive cancers	T_3M_{p+}
Orsoni et al. [24]	5 digestive cancers	T_3M_{p+}

Some particularities changed from one team to another (table 2). Fujimoto et al. [19–21] and Gilly et al. [15, 16] realized preliminary hypothermia in order to protect the brain and to avoid any deleterious increases in core temperature. The circuit could be opened (with constantly renewed heated perfusate) or closed (reusing the same perfusate).

For Koga et al. [3, 17], Fujimoto et al. [19–21] and Gilly et al. [15, 16], the abdominal wound was closed up before performing IPCH, while for Fujimura et al. [22] and Yonemura et al. [23] the wound remained opened and IPCH was realized by the means of a peritoneal expander.

Yonemura et al. [23] often performed second-look surgery (16 of 41 patients) followed by iterative chemohyperthermia (13 of 16 patients).

Thermosensitizers could be added to chemotherapy: misonidazole [20] or metronidazole [15].

Temperatures (table 3)

The temperatures reached were similar for the different teams, as the purpose was to be efficacious on tumoral development without any deleterious effect for the patient. An inflow temperature reaching 48.7 °C did not lead to any burn and allowed an intraperitoneal temperature of up to 42 °C (effective temperature against tumoral growth and local cancerous diffusion).

Heat Duration (table 2)

Heat durations varied in the different teams from 40 to 180 min. It has been proven in vitro that cancerous cells

Table 2. Design features

	Koga et al. [17]	Fujimoto et al. [19–21]	Yonemura et al. [23]	Gilly et al. [15, 16]
Hypothermia	No	30–31 °C	No	32–33 °C
Circuit	Opened	Opened	Opened	Closed
Abdominal wall	Closed	Closed	Opened	Closed
Peritoneal expander	0	0	Yes	0
Second look	0	0	Yes	0
Iterative IPCH	0	0	Yes	0
Prophylaxis	Yes	0	0	0
Thermosensitizer	0	Misonidazole	0	Metronidazole
Heat duration, min	60	90–180	40–60	90

escaped from heat action with time. Actually a medium heat duration from 60 to 90 min is used by the different teams.

Chemotherapy (table 4)

Two anticancerous drugs have been used, mitomycin C (MMC) or/and cisplatyl (CCP). Doses were similar and a combination of MMC and CCP has also been used during IPCH [23].

Indications (table 5)

The indications for IPCH were more often gastric cancer with a histologically proven peritoneal carcinomatosis. However, Gilly et al. [15, 16] performed IPCH in 3 cases for pancreatic tumors and in 3 cases for peritoneal mesothelioma.

But as shown in table 2 the main difference lies in the fact that Koga et al. [17] and Kaibara et al. [18] performed chemohyperthermia as an adjuvant therapy in peritoneal seeding-free gastric cancers with serosal invasion.

Biological Effects

Biological side effects induced by IPCH were similar in all teams, with a temporary increase in hepatic enzymes which returned to normal in 2 weeks. An early decrease in serum proteins has also been observed, it is probably due to the concomitant peritoneal dialysis and made the perfusion of albuminous concentrates necessary. The thrombocyte count decreased from the first postoperative day and returned to the normal range within 2 weeks. This phenomenon could be related to the action of anticancerous chemotherapy. Finally a transient increase in white blood cells is noted, as in any surgical process.

Table 3. Temperatures (°C)

	Koga et al. [17]	Fujimoto et al. [19–21]	Yonemura et al. [23]	Gilly et al. [15, 16]
Inflow	44–45	44.7–48.7		46–48
Outflow	40–42	43.6–46.1	42	39.42
Peritoneal		41–43		40–42
Core temperature	37.9	<41	38	39.5

Table 4. Anticancerous chemotherapy

	Koga et al. [17]	Fujimoto et al. [19–21]	Yonemura et al. [23]	Gilly et al. [15, 16]
MMC, mg/l	5	10–40	5	10
CCP, mg/l		30	20	
MMC+CCP, mg/l			5+10	

MMC = Mitomycin C; CCP = cisplatyl.

Table 5. Indications (number of patients)

	Koga et al. [17]	Fujimoto et al. [19–21]	Yonemura et al. [23]	Gilly et al. [15, 16]
Gastric cancer (prophylaxis)	6	0	0	0
Gastric cancer	0	15	53	15
Pancreatic cancer	0	0	0	3
Peritoneal mesothelioma	0	0	0	3

Table 6. Clinical complications

	Koga et al. [17] (n = 64)	Fujimoto et al. [19–21] (n = 31)	Yonemura et al. [23] (n = 41)	Gilly et al. [15, 16] (n = 23)
Anastomotic leak	5	2	1	1
Bone marrow suppression			2	
Renal failure			2	
Adhesive ileus		3		

Mortality and Morbidity (table 6)

The most frequent complications were anastomotic leak and adhesive ileus, but the different studies comparing classical surgery and surgery coupled with IPCH did not show any difference [17, 18].

2 bone marrow suppressions and 2 transitory renal failures have been observed which could be related to the additional side effects of MMC and CCP used together in these cases [22, 23].

Results

The most favorable and objective result obtained with IPCH in peritoneal seedings was the drying up of ascites (table 7). Survival rates are difficult to analyze, but they appear to be encouraging (table 8). Results obtained with adjuvant IPCH in gastric cancers with serosal invasion are impressive.

Conclusion

IPCH appears to be a safe and reliable method for the treatment of digestive cancers with peritoneal seedings. In the near future, it would be of interest to evaluate IPCH in gynecological malignancies, particularly ovarian cancers because of their frequent peritoneal dissemination and the possible enhancement of chemosensitivity. The same heating technique has already been applied in the treatment of other cancerous locations: in rectal cancers [25, 26], and in pleural seedings or mesothelioma [27].

Furthermore, the work of Koga et al. [3, 17] enables the consideration of an adjuvant use of IPCH in gastric cancers without peritoneal metastasis, and should be evaluated.

The actual protocols must be detailed (iterative IPCH, heating duration, new anticancerous drugs) and completed with other classical therapies, such as radiation therapy or immunotherapy.

Table 7. Dried up ascites

	Fujimoto et al. [19–21]	Yonemura et al. [23]	Gilly et al. [15, 16]
Ascites	9	9	10
Dried up ascites	9	7	9

Table 8. Survival rates (%)

Length of survival years	Koga et al. [17] (adjuvant)	Yonemura et al. [23]	Gilly et al. [15, 16]
1	95	55	50.4
2	78	35	47
3	73	28.5	
5	71.5		

The pharmacological modifications and the general effects induced by hyperthermia have to be investigated. What about its action on the human hormonal and immune system? Research is actually being carried out and preliminary results have been published [28–30].

Lastly, thermal dosimetry must be improved to check more precisely the temperature reached in every part of the peritoneal cavity and inside the tumor [31].

References

1 Chu DZJ, Lang NP, Thompson C, Osteen PK, Westbrook KC: Peritoneal carcinomatosis in nongynecologic malignancy. A prospective study of prognostic factors. Cancer 1989;63: 364–367.
2 Litsuka Y, Kaneshima S, Tanida O, Takeuchi T, Koga S: Intraperitoneal free cancer cells and their viability in gastric cancer. Cancer 1979; 44:1476–1480.
3 Koga S, Kaibara N, Litsuka Y: Prognostic significance of intraperitoneal free cancer cells in gastric cancer patients. J Cancer Res Chir Oncol 1984;104:236–237.
4 Vaupel P, Kallinowski F: Physiological effects of hyperthermia. Recent Results Cancer Res 1987;104:71–109.
5 Streffer C, Van Beuningen D: The biological basis for tumor therapy by hyperthermia and radiation. Recent Results Cancer Res 1987; 104:24–69.
6 Sugahara T, Watanabe M: Are malignant cells more heat sensitive than normal cells? 14th Int Symp on Clinical Hyperthermia, Dubna, USSR, 1991.
7 Turano C, Ferraro A, Strom R: The biochemical mechanism of selective heat sensitivity of cancer cells. Eur J Cancer 1970;6:67–72.
8 Gerweck LE: Hyperthermia in cancer therapy: The biological basis and unresolved questions. Cancer Res 1985;45:3408–3414.
9 Dewhirst M, Gross JF, Sim D, Arnold P, Boyer D: The effect of rate of heating or cooling prior to heating on tumor and normal tissue microcirculary blood flow. Biorheology 1984;21: 539–558.
10 Engelhardt R: Hyperthermia and drugs. Recent Results Cancer Res 1987;104:137–203.
11 Teicher BA, Kowal CD, Kennedy KA, Sartorelli AC: Enhancement by hyperthermia of the in vitro cytotoxicity of mytomycin C toward hypoxic tumor cells. Cancer Res 1981;41: 1096–1099.
12 Spratt JS, Adcock RA, Shirrell W, Trevathan S: Hyperthermic peritoneal perfusion system in canines. Cancer Res 1980;40:253.
13 Spratt JS, Adcock RA, Muscovin M, Shirrell W, McKeown J: Clinical delivery system for intraperitoneal hyperthermic chemotherapy. Cancer Res 1980;40:256–260.
14 Spratt JS, Edwards M, Kubota T, Linberg R, Tseng MT: Peritoneal carcinomatosis: Anatomy, physiology, diagnosis, management. Curr Probl Cancer 1986;10:558–584.
15 Gilly FN, Sayag AC, Carry PY, Braillon G, James I, Vollocha A, Panteix G: Intraperitoneal chemohyperthermia: A new therapy in the treatment of peritoneal seedings. Int Surg 1991; 76:164–167.
16 Gilly FN, Carry PY, Sayag AC, Braillon G, Panteix G, James I, Chatelard P: Adjuvant intraperitoneal chemohyperthermia with mitomycin C in peritoneal seedings. Jpn J Hyperthermia Oncol 1991;7:208–218.
17 Koga S, Hamazoe R, Maeta M, Shimizu N, Murakami A, Wakatsuki T: Prophylactic therapy for peritoneal recurrence of gastric cancer by continuous hyperthermic peritoneal perfusion with mitomycin C. Cancer 1988;61:232–237.
18 Kaibara N, Hamazoe R, Litsuka Y, Maeta M, Koga S: Hyperthermic peritoneal perfusion combined with anticancer chemotherapy as prophylactic treatment of peritoneal recurrence of gastric cancer. Hepatogastroenterology 1989;36:75–78.
19 Fujimoto S, Shrestha RD, Kokubun M, Ohta M, Takahashi M, Kobayashi K, Kiuchi S, Okui K, Miyoshi T, Arimizu N, Takimizawa H: Intraperitoneal hyperthermic perfusion combined with surgery effective for gastric cancer patients with peritoneal seedings. Ann Surg 1988;208:36–41.
20 Fujimoto S, Shrestha RD, Kokubun M, Kobayashi K, Kiuchi S, Konno C, Ohta M, Takahashi M: Clinical efficacy of combined therapy of intraperitoneal hyperthermochemotherapy and surgery for far advanced gastric cancer. 14th Int Symp on Clinical Hyperthermia, Dubna, USSR, 1991.
21 Fujimoto S, Shrestha RD, Kokubun M, et al: Clinical trial with surgery and intraperitoneal hyperthermic perfusion for peritoneal recurrence of gastrointestinal cancer. Cancer 1989; 64:154–160.
22 Fujimura T, Yonemura Y, Fushida S, Urade M, Takegawa S, Kamata T, Sugiyama K, Hasegawa H, Katayama K, Miwa K, Miyazaki I: Continuous hyperthermic peritoneal perfusion for the treatment of peritoneal dissemination in gastric cancers and subsequent second-look operation. Cancer 1990;65:65–71.
23 Yonemura Y, Fujimura T, Fushida S, Takegawa S, Kamata T, Katayama K, Kosaka T, Yamaguchi A, Miwa K, Miyazaki I: Hyperthermochemotherapy combined with cytoreductive surgery for the treatment of gastric cancer with peritoneal dissemination. World J Surg 1991;15:530–536.
24 Orsoni P, Soler MJ, Touchet J, Ortega D, Picaud R: Intraperitoneal hyperthermic perfusion for far advanced digestive cancer with peritoneal seeding. Study of per and post operative immediate problems. Preliminary results. 15th Int Symp on Clinical Hyperthermia, Lyon, 1992.
25 Takahashi M, Fujimoto S, Takai M, Endoh F, Oono K, Masuda Y, Shrestha RD, Kobayashi K, Kiuchi S, Kokubun M, Konno C, Obata G: Intraoperative pelvic hyperthermochemotherapy for patients with Dukes' C rectal. 15th Int Symp on Clinical Hyperthermia, Lyon, 1992.
26 Fujimoto S, Takahashi M, Endoh F, Shrestha RD, Kokubun M, Takai M, Okui K: A clinical pilot study combining surgery with intraoperative pelvic hyperthermochemotherapy to prevent the local recurrence of rectal cancer. Ann Surg 1991;213:43–47.
27 Mulsant P, Gilly FN, Carry PY, Sayag AC, Perdrix JP, Rochette C, Panteix G, Brachet A, Guibert B, Banssillon V, Bienvenu J, Braillon G: Intrapleural chemohyperthermia in the treatment of pleural malignancies: Preliminary report. 15th Int Symp on Clinical Hyperthermia, Lyon, 1992.
28 Kappel M: Effects of hyperthermia in vivo and in vitro on the activity of blood mononuclear cells. 15th Int Symp on Clinical Hyperthermia, Lyon, 1992.
29 Di Palma M, Jasmin C: Hyperthermia and tumor growth factors towards a better understanding of hyperthermia. 15th Int Symp on Clinical Hyperthermia, Lyon, 1992.
30 Coulon L, Bienvenu J, Barbier M, Gutowski MC, Panteix G, Gilly FN, Carry PY, Sayag AC, Braillon G: Cytokine production during intraperitoneal chimiohyperthermia. 15th Int Symp on Clinical Hyperthermia, Lyon, 1992.
31 Gautherie M: 3-Dimensional patient, power deposition and heat transfer modeling as basic requirement for effective clinical treatment planning in oncologic hyperthermia. 15th Int Symp on Clinical Hyperthermia, Lyon, 1992.

Oncology 1993;50:338–343

Shigeru Fujimoto
Makoto Takahashi
Kokuriki Kobayashi
Masanobu Kure
Takaaki Mutou
Hiroshi Masaoka
Haruo Ohkubo

Social Insurance Funabashi Central Hospital, Funabashi, Japan

Relation between Clinical and Histologic Outcome of Intraperitoneal Hyperthermic Perfusion for Patients with Gastric Cancer and Peritoneal Metastasis

Key Words
Gastric cancer
Peritoneal dissemination
Hyperthermic perfusion

Abstract
In 15 patients with refractory gastric cancer and peritoneal metastasis, the relationship among histologic outcome of intraperitoneal hyperthermic perfusion (IPHP), extent of peritoneal metastasis, and cause of death was studied. The IPHP treatment did not kill all the gastric cancer cells which had penetrated deeply into subperitoneal layers. In contrast, gastric cancer cells in the abdominal effusion and/or lavage vanished. Three patients with peritoneal metastasis, deeply invaded and spread all over the peritoneum, died of peritoneal recurrence and 1 died of pericardial metastasis. In the remaining 11 patients with a lesser extent of peritoneal metastasis, 5 are living without recurrence and among the other 6, 4 died of recurrence of the disease and 1 died of other causes.

Introduction

While peritoneal seeding from gastric cancer is an irreversible and fatal process, intraperitoneal chemotherapy has been reported to lead to a favorable prognosis in patients with a small volume of cancerous tissues restricted to the peritoneal surface [1]. In vitro studies [2] indicate that concentrations of drugs retained in the pelvic cavity [3] are adequate to kill colorectal cancer cell lines. Fujimura et al. [4] and Fujimoto et al. [5–7] reported treatment protocols comprising the combined efficacy of aggressive surgery for advanced gastric cancer and enhanced tumoricidal activity of hyperthermia, based on the sustained concentration of antitumor drugs in the abdominal cavity.

We reported here our findings in 15 patients with peritoneal metastasis from gastric cancer, all of whom had been treated with intraperitoneal hyperthermic perfusion (IPHP) at the time of aggressive surgery for primary foci. Tissue and effusion specimens were obtained before and after IPHP. Morphologic effects on the cancer cells and the cytohistologic findings were compared between pre- and post-IPHP samples. The relationship among morphologic effects of the IPHP, the extent of peritoneal metastasis and cause of death is discussed.

Materials and Methods

Fifteen patients included in this report had advanced gastric cancer with peritoneal metastasis and/or peritoneal effusion. At laparotomy, careful inspection both for the primary lesion and for metastatic foci on the peritoneo-serosal surface was carried out in all quadrants of the abdominal cavity. In the 15 cases with a resectable primary lesion, peritoneo-serosal metastatic foci were harvested and

The authors thank M. Ohara for valuable comments.

Shigeru Fujimoto
Social Insurance Funabashi Central Hospital
6-13-10, Kaijin
Funabashi 273 (Japan)

0030–2414/93/0505–0338
$2.75/0

these tissues and/or effusion harvested were compared with the post-IPHP specimens.

For these 15 patients, aggressive surgery such as total gastrectomy combined with excision to remove residual foci in adjacent organs was done. The survival rates were examined and the relationship between the antitumor effects of the IPHP and the cause of death was analyzed.

Cytohistologic evaluations were made based on histologic changes between pre- and post-IPHP cancer cells in the peritoneal tissues and in ascitic effusion and/or peritoneal lavage. Changes included nuclear pyknosis, karyorrhexis, and karyolysis as well as shrinkage of cells [8, 9]. In 6 cases with specimens from ascitic effusion and peritoneal lavage, the presence or absence of cancer cells in the repeated post-IPHP peritoneal lavages served as another reliable criterion.

Cytohistologic findings were recorded as 3+, when there were no cancer cells in the peritoneal lavage and all nuclei of the cancer cells under the peritoneum were pyknotic or karyolytic. When a great majority of the nuclei were pyknotic or karyolytic, 2+ findings was recorded. When cancer cells were detected in the deep subperitoneal layer, though not just beneath the peritoneum, as well as when only a few cancer cells were present in repeated peritoneal lavages, they were recorded as being 1+.

With respect to the clinical classification of peritoneal metastasis as outlined by the Japanese Research Society for Gastric Cancer [10], disseminating peritoneal metastasis was classified into the following 3 groups, that is, p1, p2, and p3. In brief, p1 is a metastasis disseminating to the adjacent peritoneum (above the transverse colon), including the greater omentum. p2 is several scattered metastases to the distant peritoneum, including the peritoneum below the transverse colon and the abdominal surface of the diaphragm. Numerous disseminating metastases evident all over the peritoneum are group p3. For one with a positive cytology in the peritoneal lavage, with no visible peritoneal metastasis, a p1 was designated.

Intraperitoneal hyperthermic perfusion (IPHP) using 10.3 ± 1.0 μg/ml of mitomycin C (MMC) was carried out immediately after the surgery. A detailed description of the IPHP treatment has been reported elsewhere [5–7]. Briefly, under hypothermic general anesthesia to prevent a rise in body temperature, IPHP was performed with a closed peritoneal perfusion using 3,000–4,000 ml of a perfusate containing MMC. Temperatures of the perfusate at the inflow and outflow points were kept at 45.0–46.2 °C and 43.5–45.1 °C, respectively (fig. 1). Before the onset of IPHP, the systemic temperature was lowered to 31 °C using a cooling mat and ice-bags (fig. 1). The perfusion time was 118 ± 14 min (table 1) and at the end of the IPHP, the pulmonary artery temperature measured by a Swan-Ganz catheter did not exceed 40 °C. Throughout the IPHP, the temperature was measured at the following points: the inflow and outflow points, Douglas' and Morison's pouches, and the pulmonary artery.

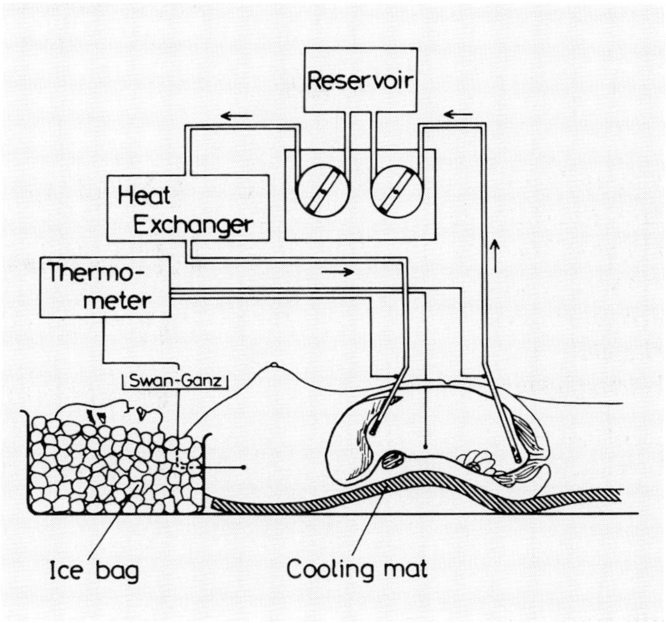

Fig. 1. Schematic drawing of the sagittal section of the abdominal cavity in case of intraperitoneal hyperthermic perfusion (IPHP). Arrows show flow direction of the perfusate.

Results

Peritoneo-serosal tissues were available post-IPHP for only 9 patients, because of complete disappearance in 6 patients of the disseminating foci on the peritoneo-serosal surface (table 1). In these 9, the disseminating foci on the peritoneo-serosal surface diminished in size and height after IPHP to about half the pre-IPHP size, even when allowances were made for IPHP-related edema.

The peritoneal effusion and intra-abdominal lavage were studied in all patients pre- and post-IPHP. Although the peritoneal effusion in all 15 patients did contain numerous cancer cells pre-IPHP (fig. 2A), post-IPHP there were no cancer cells in the exudate from Douglas' pouch (fig. 2B).

Figure 3 shows histologic changes in the peritoneal surface and in gastric cancer cells, as induced by the IPHP treatment. At the area just beneath the peritoneal surface, changes were most remarkable. All nuclei of the cancer cells were pyknotic or karyolytic. There was every evidence that cells in the cancer nests disappeared with antitumor activity of the IPHP.

Figure 4 shows histologic changes evident of the IPHP treatment. The extent of tumoricidal activity of the IPHP does have limitations; we estimate the range of cytocidal effect of the IPHP to be about 5 mm deep from the peritoneo-serosal surface, when IPHP treatment is given as described above.

With respect to the relation among extent of peritoneal metastasis, antitumor effects of the IPHP and cause of death (table 2), 3 of 4 patients with p1 or p2 and histologic findings of 2+ or 3+ survive without any recurrence for 39.0 ± 3.6 months after the IPHP, and 2 of 4 patients

Table 1. Clinical data on 15 patients

Patient No.	Age/ sex	Clinicopathologic findings			Surgery	IPHP			Cytohisto-logic findings after IPHP effusion/ peritoneum	Clinical status (cause of death)	Follow-up months
		peritoneal metastasis	primary tumor			mitomycin C		perfusion time min			
			extent of invasion	histol-ogy		total dose mg	concen-tration at the start, μg/ml				
1	58/M	p1	T4	D	PD	40	10	90	3+/ND	NED	48
2	40/M	p1	T4	UD	TG, Spl and TP	30	10	150	3+/2+	NED	42
3	52/M	p1	T3	UD	STG	35	10	120	3+/ND	NED	28
4	63/M	p1	T3	UD	TG, Spl, TBP and Trc	35	10	120	3+/ND	DOD (IAT)	8
5	46/F	p1	T4	D	STG and Chol	50	14	120	3+/ND	DOC	7
6	70/M	p2 (Asc: 1,400)	T4	UD	TG, Spl and BTP	40	10	90	3+/ND	DOD (HM)	10
7	30/F	p2	T3	UD	TG, TP and Oo	30	10	120	3+/3+	NED	34
8	53/F	p2	T4	D	TG, Spl and Oo	30	10	120	3+/3+	NED	41
9	38/F	p2	T3	UD	TG, Spl, TP and Oo	30	10	120	3+/ND	DOD (HM)	12
10	39/F	p2	T3	D	TG, Spl, BTP, Chol and Oo	30	10	120	3+/1+	DOD (Ple)	8
11	37/F	p2	T3	UD	TG and Spl	30	10	120	3+/3+	DOD (HM)	5
12	33/F	p3 (Asc: 4,500)	T3	UD	TG, Spl, Oo and rHc	30	10	130	3+/2+	DOD (IAT)	18
13	51/F	p3 (Asc: 3,500)	T3	D	TG and Spl	40	10	120	3+/2+	DOD (Pec)	30
14	39/M	p3 (Asc: 4,000)	T4	UD	TG and Spl	30	10	120	3+/1+	DOD (Per)	9
15	62/F	p3 (Asc: 2,500)	T4	UD	TG, Spl, TBP and Trc	26	10	120	3+/2+	DOD (Per)	49

UD = Undifferentiated adenocarcinoma; D = differentiated adenocarcinoma; Asc = volume (ml) of ascitic effusion; TG = total gastrectomy; Spl = splenectomy; Oo = oophorectomy; rHc = right hemicolectomy; TP = tail pancreatectomy; BTP = body and tail pancreatectomy; Chol = cholecystectomy; PD = pancreaticoduodenectomy; Trc = transverse colectomy; ND = not done; DOD = died of disease; DOC = died of other cause; NED = no evidence of disease; IAT = intra-abdominal recurrent tumors; Pec = pericardial metastasis; Ple = pleural metastasis; Per = peritoneal metastasis; HM = hepatic metastasis.

with p1, in whom there was a complete disappearance of the peritoneal matastases following IPHP, are also living at 38 months following treatment. In contrast, in 4 patients with p3 with histologic findings of either 1+ or 2+, 3 died of peritoneal metastasis.

Discussion

Although the antitumor efficacy of the IPHP was apparent in tissues just beneath the peritoneum, there were no effects on the deep layer in the peritoneal tissues. In gastric cancer cells present in the peritoneal effusion, decisive cytologic changes were detected just after IPHP treatments; activities attributed to the direct cytotoxicity of the IPHP. Nevertheless, hyperthermia plus antitumor drug lead to only a limited penetration through peritoneal tissues.

Link [11] reported in vitro antitumor effects of MMC as a function of drug concentration and exposure time, using a cell line from human colon carcinoma and a human tumor colony assay with soft agar. The lethal drug concentration of MMC was 1.0 μg/ml for a 6-hour exposure and the concentration for 50% growth inhibition was 0.3 μg/ml for a 60-min exposure. Park et al. [2] also did in vitro chemosensitivity tests on 11 different cell lines from human colorectal cancer. When MMC was used for a 96-hour exposure, the mean value of the 50% inhibitory dose was 0.45 μg/ml. In our patients, MMC concentrations in the IPHP treatment for over 2 h were approximately 3 μg/ml on average [7]. Based on these findings, MMC concentrations during the IPHP might be inadequate in case of cancer cells with a low sensitivity to MMC.

Wallner et al. [12] found, in in vitro studies on 3 different Chinese hamster ovary cell lines, that the cytotoxicity with 2 μg/ml MMC at 37°C for 1 h was 3×10^{-2} in the

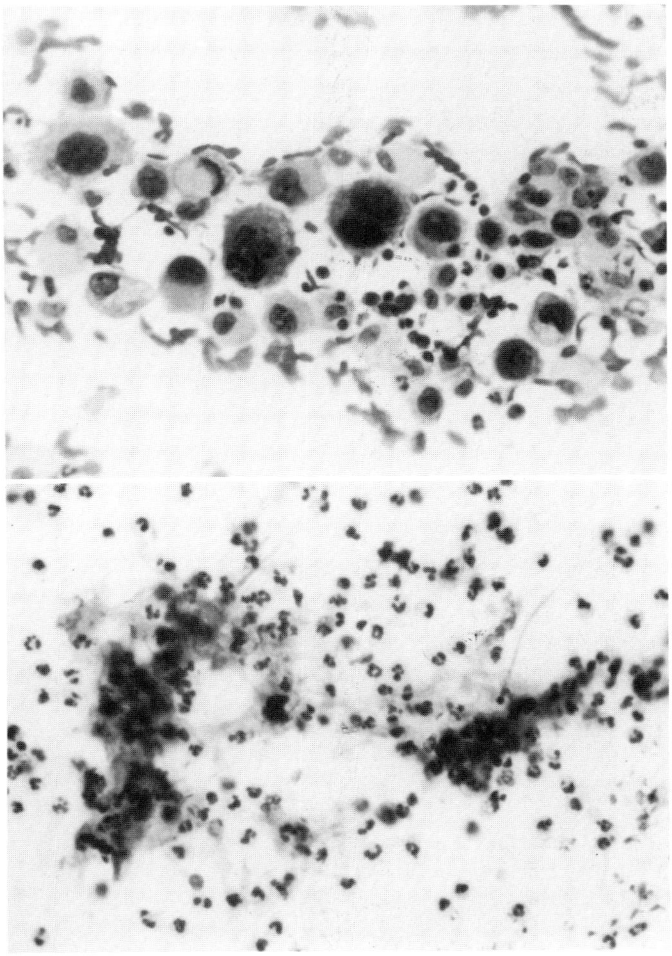

2A

2B

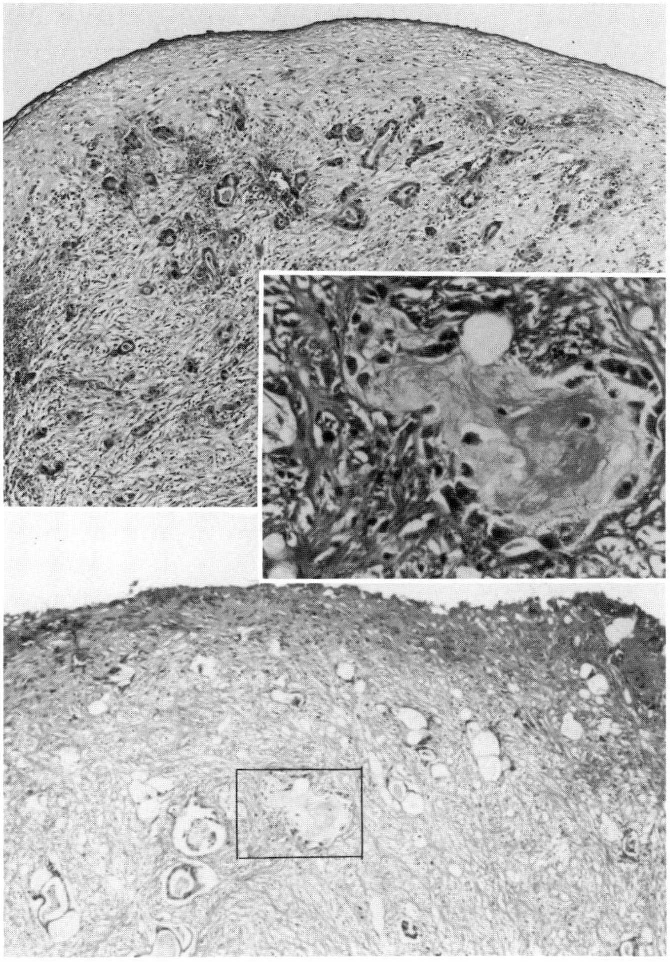

3A

3B

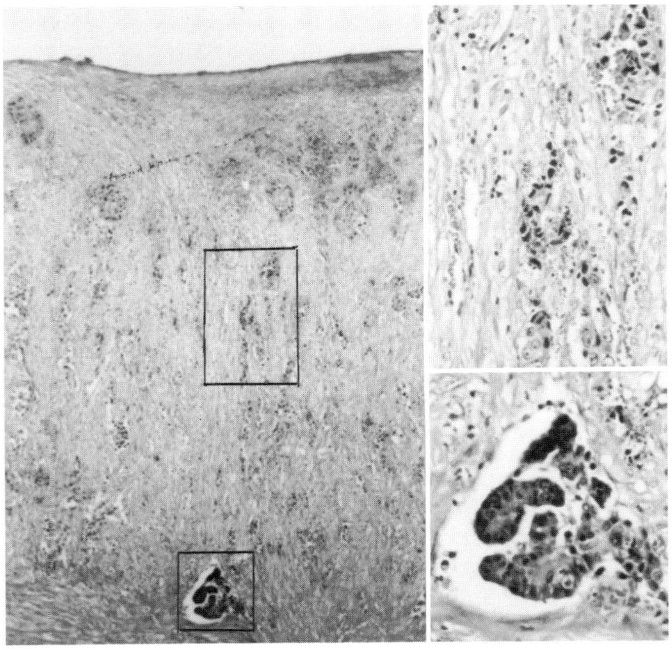

4

Fig. 2. Gastric cancer cells in peritoneal effusion or lavage. **A** Pretreatment specimen from patient 14. Note the many gastric cancer cells in cluster form. Papanicolaou. × 200. **B** Tumor cells in patient 6 surrounded with neutrophils 96 h after IPHP. Papanicolaou. × 200.

Fig. 3. Cytohistologic changes in the peritoneum and gastric cancer cells in the subperitoneal layer. **A** Pre-IPHP peritoneum with gastric cancer cells. HE. × 40. **B** Post-IPHP specimen. HE. × 40. Higher magnification (boxed area) illustrating numerous pyknotic and karyolytic cancer cells. HE. × 200; in the corners of this higher magnification it can be seen that the cancer cells vanished but adenomatous structures are evident.

Fig. 4. Histologic changes following IPHP treatment in patient 13. HE. × 40. Higher magnification of boxed areas showing damage to cancer cell nests and differentiated adenocarcinoma with lymphatic plugging. HE. × 200.

Table 2. Relationship among extent of peritoneal metastasis, efficacy of the IPHP and cause of death

Peritoneal metastasis	Number of patients	Post-IPHP histologic findings of peritoneum		NED	DOD: peritoneal metastasis	DOD: hepatic metastasis	DOD: pleural matastasis	DOD: pericardial metastasis	DOC
P1	5	ND	4	2 (48,28)	1 (8)				1 (7)
		2+	1	1 (42)					
P2	6	ND	2			2 (10,12)			
		1+	1				1 (8)		
		3+	3	2 (34,41)		1 (5)			
P3	4	1+	1		1 (9)				
		2+	3		2 (18,49)			1 (30)	

ND = Not done; NED = no evidence of disease; DOD = died of disease; DOC = died of other cause; number in parentheses = months after IPHP.

survival fraction in cells sensitive to MMC, whereas at 43.5°C, it was 1.1×10^{-3}. Their report was based on the thesis that MMC-resistant cells exposed to MMC combined with heat might enhance the response to the drug.

Although the enhanced antitumor efficacy of the IPHP was seen in cancer cells present in abdominal effusion as well as on and just beneath the peritoneum, figure 4 shows that the hyperthermic tumoricidal efficacy combined with MMC does not permeate deeply enough through the peritoneal tissues. In the case of hyperthermic perfusion for cancerous foci beneath the peritoneal surface, unlike an external hyperthermia for malignant solid tumors, the normal tissue surrounds the deeply invaded cancerous foci. Under such circumstances, heat spreads through the blood stream in the vessels dilated in proportion to temperatures and, consequently, focal high temperature on the peritoneal surface drops in the tissue where those foci locate.

Sugarbaker et al. [13] reported findings in cases of postoperative intraperitoneal chemotherapy with 5-FU and MMC in 9 patients with peritoneal carcinomatosis from a mucinous cystadenocarcinoma of the colon or appendix. After cytoreductive surgery for removing as much as possible of the extensive mucinous ascites or bulky intra-abdominal mucinous tumors, 12 mg/m^2 of MMC and 20 mg/kg of 5-FU were given intraperitoneally in 2,000 ml of 1.5% dextrose peritoneal dialysis solution, under the following schedule. MMC was given on alternate days and 5-FU for 5 consecutive days and this cycle was repeated 5.2 times on average. This intraperitoneal chemotherapy led to a positive clinical outcome in 6 of the 9 patients and the histologic changes in mucinous tumors, as a result of a direct cytotoxicity of both drugs, were remarkable on the peritoneal surface.

In our 9 patients from whom posttreatment peritoneal tissues were harvested, 3 died of peritoneal recurrence or intraperitoneal recurrent tumors; in these patients, residual microscopic foci were present in regions distant from the peritoneal surface, as shown in figure 4. We attributed this to a decline in temperature and drug concentration due to capillary dilatation induced by hyperthermia. In contrast, in 2 of the remaining 6 patients, a distant metastasis in the liver prohibited any beneficial local effectiveness with the use of IPHP.

While there are limitations of IPHP treatment, this approach combined with extensive surgical excision does provide one method of treating gastric cancer patients with peritoneal metastasis.

References

1 Sugarbaker PH, Gianola FJ, Speyer JL, Wesley R, Barofsky I, Meyers CE: Prospective randomized trial of intravenous versus intraperitoneal 5-fluorouracil in patients with advanced primary colon or rectal cancer. Surgery 1985; 98:414–421.

2 Park J-G, Kramer BS, Steinberg SM, Carmichael J, Collins JM, Minna JD, Gazdar AF: Chemosensitivity testing of human colorectal carcinoma cell lines using a tetrazolium-based colorimetric assay. Cancer Res 1987;47:5875–5879.

3 Fujimoto S, Takahashi M, Endoh F, Shrestha RD, Kokubun M, Takai M, Okui K: A clinical pilot study combining surgery with intraoperative pelvic hyperthermochemotherapy to prevent the local recurrence of rectal cancer. Ann Surg 1991;213:43–47.

4 Fujimura T, Yonemura Y, Fushida S, Urade M, Takegawa S, Kamata T, Sugiyama K, Hasegawa H, Katayama K, Miwa K, Miyazaki I: Continuous hyperthermic peritoneal perfusion for the treatment of peritoneal dissemination in gastric cancers and subsequent second-look operation. Cancer 1990;65:65–71.

5 Fujimoto S, Shrestha RD, Kokubun M, Ohta M, Takahashi M, Kobayashi K, Kiuchi S, Okui K: Intraperitoneal hyperthermic perfusion combined with surgery effective for gastric cancer patients with peritoneal seeding. Ann Surg 1988;208:36–41.

6 Fujimoto S, Shrestha RD, Kokubun M, Kobayashi K, Kiuchi S, Takahashi M, Konno C, Ohta M, Koike S, Kitsukawa Y, Mizutani M, Okui K: Clinical trial with surgery and intraperitoneal hyperthermic perfusion for peritoneal recurrence of gastrointestinal cancer. Cancer 1989;64:154–160.

7 Fujimoto S, Shrestha RD, Kokubun M, Kobayashi K, Kiuchi S, Konno C, Takahashi M, Okui K: Pharmacokinetic analysis of mitomycin C for intraperitoneal hyperthermic perfusion in patients with far-advanced or recurrent gastric cancer. Reg Cancer Treat 1989;2:198–202.

8 Goudie RB: Cell damage; in Anderson JR (ed): Muir's Textbook of Pathology, ed 11. London, Edward Arnold, 1980, pp 7–42.

9 Scarpelli DG, Chiga M: Cell injury and errors of metabolism; in Anderson WAD, Kissane JM (eds): Pathology, ed 7. Saint Louis, Mosby, 1977, pp 90–147.

10 Japanese Research Society for Gastric Cancer: The general rules for the gastric cancer study in surgery and pathology. Jpn J Surg 1981;11: 127–139.

11 Link KH: Basic concepts for the application of mitomycin C in regional cancer treatment; in Taguchi T, Aigner KR (eds): Mitomycin C in Cancer Chemotherapy Today. Tokyo, Excerpta Medica, 1991, pp 62–71.

12 Wallner KE, Banda M, Li GC: Hyperthermic enhancement of cell killing by mitomycin C in mitomycin C-resistant Chinese hamster ovary cells. Cancer Res 1987;47:1308–1312.

13 Sugarbaker PH, Landy D, Jaffe G, Pascal R: Histologic changes induced by intraperitoneal chemotherapy with 5-fluorouracil and mitomycin C in patients with peritoneal carcinomatosis from cystadenocarcinoma of the colon and appendix. Cancer 1990;65:1495–1501.

Oncology 1993;50:344–347

Igor A. Mikhalkin
Zakhar G. Iashvili
Vladimir L. Bykov

Division of Oncology, Kiev Research Institute of Otolaryngology, Kiev, Ukraine

Thermoradiotherapy Combined with a Proteolysis Inhibitor (Contrical) in the Treatment of Head and Neck Cancer

Key Words
Hyperthermia
Adjuvant therapy
Interferon
Contrical
Tumor

Abstract
For the prophylaxis of head and neck cancer recurrence and metastases we used radiation therapy, microwave hyperthermia and adjuvant therapy (inhibitor of proteolysis contrical and α-interferon). One hundred forty patients with primary cancer were treated. After 1 year of observation time and absence of tumor the patients were examined once every 3 months. The use of contrical and interferon combined with radiation therapy and microwave hyperthermia increased the 3-year survival rate by 37%, the number of patients without recurrence by 39% and decreased the appearance of regional metastases to 8.4%. Adjuvant therapy is discussed.

Introduction

Despite the high success rates in cancer treatment, the majority of patients die of recurrence or metastases [1–3]. The frequency of regional and distal metastases of head and neck cancer is approximately 36–57%. Therefore, the carrying out of methods and finding of prophylactic treatment to prevent recurrence or metastases is a very current problem in oncology. Unfortunately, there have not been many clinical investigations in this area. The aim of the present study is the clinical investigation for a possible prophylaxis against recurrence or metastases after radiation therapy combined with hyperthermia in patients with head and neck cancer. As prophylaxis, we chose a natural proteolysis inhibitor – contrical. The premise for using contrical was data in the literature stating that natural polyfunctional inhibitors of polypeptide proteolysis which derive from the pancreas, parotid gland or lungs of cattle include contrical, a substance which was able to inhibit the evasiveness of malignant cells [4, 5] and decrease the growth and spread of tumors in experimental animals [6–9]. Injection of a proteolysis inhibitor increases the length of survival [8], with tumor regression [10] and decrease in metastases [11]. Some authors believe that the actions of the proteolysis inhibitors are accomplished through the immune system [6, 12, 13].

Materials and Methods

Seventy-four patients with primary head and neck tumors were treated at the Division of Oncology, Kiev Research Institute of Otolaryngology. All patients were divided into three groups: group 1: 25 patients who underwent radiation therapy; group 2: 25 patients who underwent radiation therapy combined with hyperthermia; group 3: 24 patients who underwent radiation therapy combined with hyperthermia and the proteolysis inhibitor contrical.

Igor A. Mikhalkin, MD, PhD
Gogolevska Str. 34–24
Kiev-54
Ukraine, 252054

0030–2414/93/0505–0344
$2.75/0

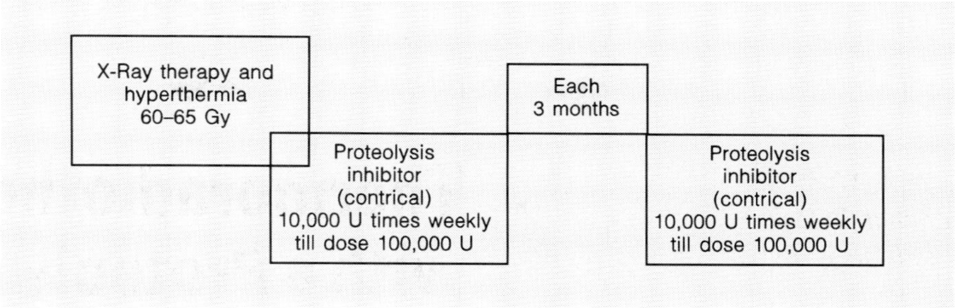

Fig. 1. Scheme of the prophylaxis against recurrence or metastasis by a proteolysis inhibitor (contrical).

Table 1. Distribution of patients according to primary tumor location

Treated patients	Tumor location									
	nasal cavity paranasal sinus		oropharynx		hypopharynx		larynx		total patients	
	n	%	n	%	n	%	n	%	n	%
X-ray therapy control group	3	12.0	7	28.0	4	16.0	11	44.0	25	100
X-ray therapy + hyperthermia	2	8.0	5	20.0	5	20.0	13	52.0	25	100
X-ray therapy + hyperthermia + proteolysis inhibitor (contrical)	1	4.2	6	25.0	5	20.8	12	50.0	24	100

Table 2. Distribution of patients according to primary tumor spread and regional metastatis

Treated patients	Tumor spread and regional mMetastases													
	T_3N_0		T_4N_0		T_3N_1		T_3N_2		T_4N_1		T_4N_2		total patients	
	n	%	n	%	n	%	n	%	n	%	n	%	n	%
X-ray therapy control group	10	40.0	2	8.0	4	16.0	2	8.0	5	20.0	2	8.0	25	100
X-ray therapy + hyperthermia	9	36.0	3	12.0	5	20.0	2	8.0	4	16.0	2	8.0	25	100
X-ray therapy + hyperthermia + proteolysis inhibitor (contrical)	9	37.5	4	16.7	5	15.6	3	20.8	2	12.5	1	4.2	24	100

The distribution of patients according to primary tumor location is represented in table 1, and according to primary tumor spread and regional metastases in table 2. All patients were selected for each group by a randomized method and their diagnoses were proven histologically. All patients have imaging modalities to confirm the diagnosis. Local microwave hyperthermia was achieved by an electromagnetic field at 915 MHz with 0.5–0.6 W/cm^2 using the external technique and 2,450 MHz with 7–8 W/cm^2 using the intracavitary technique.

Temperature control was accomplished using copper-constant thermocouples that went around the tumor and its underlying tissue. The tumor was heated up to 42.5–43°C. Hyperthermia was performed twice weekly. Tumor response and recurrence rates are histologically proven. Monthly follow-up evaluation was done by indirect laryngoscopy, i.e. flexible orifice naso-oro-pharyngo-laryngoscopy. After 1 year's observation and absence of tumor, the patients were observed once every 3 months.

Radiation therapy was given at a dose of 60–65 Gy to the primary tumor and 45–50 Gy to regional metastases in a split course with a 2-week break after a dose of 40 Gy. Injection of 10,000 U contrical was given intravenously slowly 3 times a week in 20 ml. The injections were begun after 2 weeks' break and a dose of 40 Gy. The total dose of contrical was 100,000 U. After the course of radiation alone and combined with hyperthermia, the course of injection on contrical was repeated (fig. 1) over 3–4 months about 3 times a year. Each time, the total dose of contrical was 100,000 U.

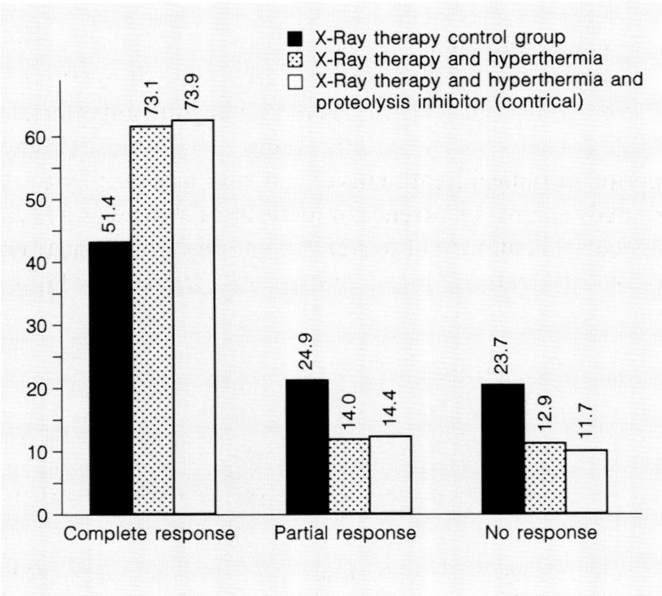

Fig. 2. Short-term response in patients with head and neck cancer.

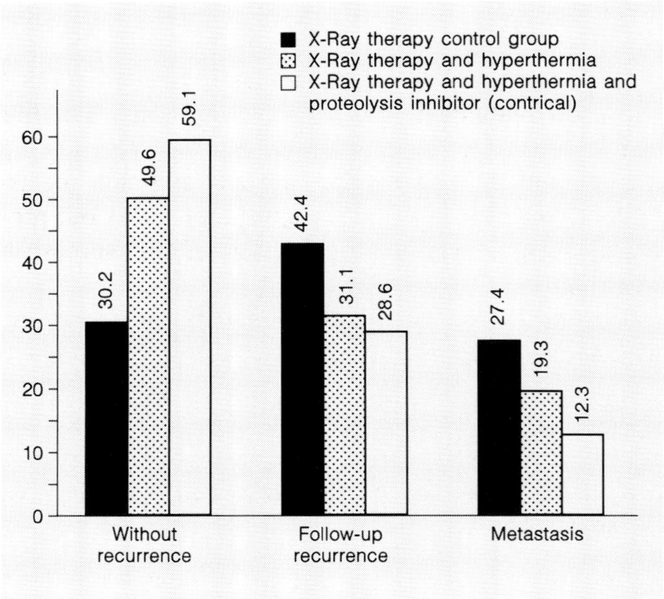

Fig. 4. Tumor recurrence: 3-year observation.

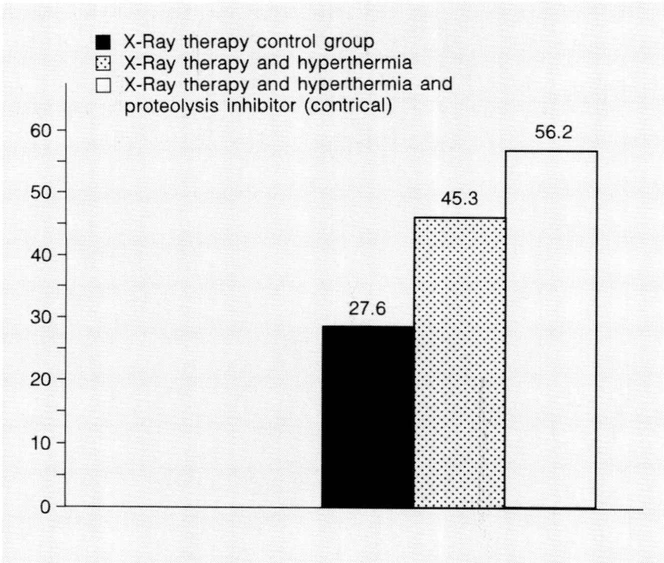

Fig. 3. Three-year survival rate.

Results

The short-term response in patients with head and neck cancer is represented in figure 2. The local hyperthermia essentially increases the complete response of head and neck cancer, the proteolysis inhibitor contrical does not have an influence on the short-term response. The 3-year survival rates are represented in figure 3. Local hyperthermia increased survival compared with radiation therapy alone to about 18%, the proteolysis inhibitor combined with radiation therapy and hyperthermia increased the 3-year survival to 30% compared with radiation therapy alone and to 11% compared with radiation therapy in combination with hyperthermia.

Tumor recurrence and metastases in our 3-year observation is presented in figure 4. Local hyperthermia increased the number of patients without recurrence in the 3-year observation to 19% compared with radiation therapy and the proteolysis inhibitor contrical increased this rate to 29% compared with radiation therapy alone and to 10% compared with radiation therapy and hyperthermia combined. We have observed regional metastases after radiation therapy alone and complete response in 27.4% of the patients, after radiation therapy combined with local hyperthermia in 19.3%, and after radiation therapy combined with local hyperthermia and proteolysis inhibitor in only 12.3%. Using contrical promoted extension of the period of remission. We observed the appearance of regional metastases or recurrence mainly in the first year after radiation therapy and no patients have recurrence or metastases in the first year's observation after radiation therapy combined with local hyperthermia and contrical.

Discussion

Surgery, radiation and chemotherapy have achieved very good treatment results in oncology patients, but after primary treatment all patients are observed without any therapy and their physicians can only ascertain metastases or recurrence. It is very important to carry out prophylaxis against recurrence or metastases after successful treatment.

Contrical in patients with a partial response or those without any response after radiation therapy did not give positive results. All patients (8) had died during the 9 months of observation. Using contrical with radiation therapy for treatment of recurrence or metastases without surgery as a palliative course has an insignificant effect.

The results testify the expediency of earlier use (in the period of radical treatment) of contrical as a prophylactic remedy against recurrence or metastases in head and neck cancer. It is important to start the prophylaxis at the time of radical treatment and continue for a long period of time (2–3 years).

References

1 Ogolzova BS: The malignant tumors of the upper respiratory tract. Med Mosc 1984:224.
2 Olshansky VO, Bituzky PG, Kovalev BN: The prophylaxis of metastases of cancer of larynx by different methods of influence on zone of regional metastases (randomized investigation). Sov Med 1985;121:38–42.
3 Pashes AI: The head and neck tumor. Med Mosc 1983:415.
4 Latner AL, Longstaff E, Pradhan K: Inhibition of malignant cell invasion in vitro by a proteinase inhibitor. Br J Cancer 1973;27:460–464.
5 Shnebli HP, Burger MM: Selective inhibition of growth of transformed cells by protease inhibitors. Proc Natl Acad Sci USA 1972;69:3825–3827.
6 Mishin YB: The influence of inhibitor of proteinase-contrical on the transplantation and growth of a solid carcinoma of Ehrlich in mice. Vop Oncol 1982;1:73–75.
7 Back N, Steger A: Effect of aprotinin, EACA and heparin on growth and vasopeptide system of Murphy-Sturm lymphosarcoma. Eur J Pharmacol 1976;38:313–319.
8 Lage A, Diaz J, Gonzales J: Effect of proteinase inhibitors in experimental tumors. Neoplasma 1978;25:257–259.
9 Thompson AW, Pugh-Humphreys RGP, Horne CHW, et al: Aprotinin and growth of Walker 256 carcinosarcoma in the rat. Br J Cancer 1977;35:454–460.
10 Latner AL, Longstaff E: Antitumor activity of aprotinin. Br J Cancer 1974;30:60–67.
11 Giraldi T, Nisi C, Sava G: Lysosomal enzyme inhibitors and antimetastatic activity in the mouse. Eur J Cancer 1977;13:1321–1323.
12 Zolotuchin NA, Revin AN: Use of inhibitors of proteinase by cancer intoxication. Surgery 1977;6:110–111.
13 Lather A, Turner G: Effect of aprotinin on immunological resistance in tumor-bearing animals. Br J Cancer 1976;33:535–538.

Oncology 1993;50:348–352

P.Y. Carry[a], A. Brachet[b]
F.N. Gilly[b], A.C. Sayag[a]
P.Y. Petit[a], C. Rochette[a]
P. Mulsant[b], B. Guibert[b]
G. Braillon[b], V. Banssillon[a]

Departments of
[a] Anaesthesia and Intensive Care, and
[b] General and Thoracic Surgery,
Centre Hospitalier Lyon-Sud, Lyon, France

A New Device for the Treatment of Pleural Malignancies: Intrapleural Chemohyperthermia Preliminary Report

Key Words
Pleural malignancies
Mitomycin C
Intrapleural therapy
Hyperthermia

Abstract
The prognosis of malignant pleural tumors remains extremely unfavorable. The aim of this study is to evaluate the combination of intrathoracic intrapleural chemotherapy and intrapleural hyperthermia (ITCH) in these diseases. Under anesthesia, 5 men were studied. After pleurectomy for mesothelioma (3/5) or adenocarcinoma (2/5), ITCH is carried out for over 60 min, either with mitomycin C (4/5) or cisplatin (1/5). No pre- or postoperative death occurred. The maximal pleural temperature is 42.6°C. The blood level of mitomycin C never reached the systemic toxic level. All the patients were discharged from the surgical ward, 3 are still alive 15 months later. Therefore, ITCH appears to be a safe and reliable therapy.

Malignant pleural effusion or mesothelioma has appeared increasingly in the last two decades [1]. Nevertheless, the prognosis is unfavorable with or without treatment (survival rate from 6 to 16 months). The surgical treatment, extrapleural pneumectomy or pleurectomy/decortication [2] remains the mainstay of treatment. Radiotherapy and/or chemotherapy can be proposed in combination with surgery [3]. Actually, intrapleural interferon [4] or chemotherapy [5] are evaluated but no standard treatment for diffuse malignant pleural effusions is proposed. We evaluated and studied the feasibility of the combination of intrathoracic intrapleural chemotherapy and intrapleural hyperthermia (ITCH) for diffusive malignant pleural effusions.

Material and Methods

The protocol was realized after informed consent of the patients and their families. The inclusion criteria were (a) patients with diffusive malignant pleural disease; (b) patients without extrathoracic metastasis; (c) acceptable cardiorespiratory function. Five men aged from 42 to 72 years, scheduled for surgery, were treated with ITCH with mitomycin (c) or cysplatin. Malignant pleural diseases were found during thoracoscopy for preoperative cytological and histological evaluations. Diagnosis, localisation of cancer, and form of surgery are presented in table 1.

The procedure was realized under general anesthesia, lateral position and one lung ventilation. A thoracotomy was performed to explore the pleural cavity, take cytological and histological samples, and for pleurectomy when possible. Before the surgical procedure, a Swan-Ganz catheter was inserted inside the pulmonary artery to monitor central temperature and hemodynamic parameters. A 'pre-

P.Y. Carry, MD
Department of Anaesthesia and Intensive Care
Centre Hospitalier Lyon-Sud
F-69310 Lyon Pierre-Bénite (France)

0030–2414/93/0505–0348
$2.75/0

Table 1. Data on patients treated by IPCH

Patient	Age years	Diagnosis	Stage	Surgery
1	71	mesothelioma	I	0
2	42	metastasis ADK	IV	pleurectomy
3	72	mesothelioma	I	pleurectomy
4	64	mesothelioma	I	pleurectomy
5	49	metastasis ADK	IV	pleurectomy

ADK = Adenocarcinoma.

Table 2. Clinical follow-up

Patient	Complication	Discharge days	Evolution	Survival months
1	0	14	hepatic metastasis	4, dead
2	aeric leak	22	increased ACE	20, alive
3	pleural clotting	29	pleural	22, alive
4	0	10	pleural/general	11, dead
5	0	10	pleural	15, alive

Table 3. Mean values of biological parameters

	CPK IU/l	Protein g/l	CRP g/l	Lactate acid m*M*
J–1	57	70	10	1.6
et	22	6.5	7.5	0.1
T 0	390	60.2	17.6	2.7
et	245.6	3.1	14.1	0.9
T 30	583.8	56	6.7	2.8
et	198.9	3.5	0.9	0.7
T 60	604.8	57.2	7.5	3.1
et	182.5	3.5	1.5	1
T 90	571.8	51.5	10.6	3.3
et	188.1	2.7	8.8	0.7
T 360	449.5	54.4	17	3
et	287.7	2.9	16.2	0.7
T 480	647.7	51.8	16.5	3.9
et	262.2	2.4	16.5	1.7
J 1	1,078.3	50	60.3	2.1
et	366.9	7.1	13.4	

et = Ecartype. The mean values of the differents parameters are measured during preoperative day J–1, at the beginning of the device T 0, after 30 min T 30, 60 min T 60, 90 min T 90, 360 min T 360, 480 min T 480, and after 1 day J1.

hyperthermia' hypothermia (under 33 °C) was realized by the mean ice-bags, cold wraps on the legs (Spenco LTD., Teynning, UK) and cooling-hat (Pinopharma, Freudenstadt, FRG). Before closure of the thoracotomy, the equipment needed for ITCH was inserted into the pleural cavity: (a) 1 silicone drainage, Inflow drainage (William Harvey, Bard Cardiopulmonary Division, USA), inserted at the top of the cavity; (b) 1 silicone drainage, Outflow drainage (William Harvey), inserted at the bottom. Thermic probes (Mallinkrodt) were also inserted: (a) in the pleural cavity; (b) 1 on the inflow and 1 on the outflow drainage (from 8 to 10 cm from the skin); (c) 1 in the pericardic space. Then thoracotomy was closed and the inflow and outflow drainages were connected to a close sterile circuit where four liters of perfusate (Travenol Laboratory) were propelled by means of a roller pump (Cobe) at the rate of 200 cm^3/min. This liquid was heated through a thermic exchanger (Dideco) connected to a heating circuit. Intra- and extrapleural thermic probes were connected to the thermic reader to be monitored each 10 min.

ITCH was carried out for over 60 min paying close attention to respiratory and hemodynamic functions, inflow temperatures were less than 45 °C. During ITCH, blood samples were taken every 30 min to determine white cell, red cell and platelet counts, coagulation, creatinine phosphokinase (CPK), protidemia, CRP, lactate acid, and hematosis. MMC levels in blood and perfusate was also measured. The different temperatures were monitored every 10 min. After ITCH, all the patients were placed in the intensive care unit where these differences were also determined during 1 day.

Results

No pre- or postoperative death occurred. No morbidity (respiratory, cardiac) was observed during the protocol. All the patients were discharged from hospital (table 2). An aeric leak occurred in a 42-year-old man (patient 2) resulting in his hospital discharge being somewhat delayed. Pleural clotting occurred in a 72-year-old man (patient 3) necessitating surgery before recovery.

The data of the monitored parameters are given in tables 3 and 4. A slight decrease of protidemia and an increase of CPK, remaining high 1 day after ITCH, were observed without a significant consequence. As during traditional surgery, an increase of white blood cells was observed without infectious significance. During the time of ITCH neither symptomatic modification of other hematologic parameters nor variation of factors of coagulation occurred.

Thermal variations are represented in figure 1. No problems were observed concerning thermal tolerance. The maximal inflow temperature observed was 48.2 °C in patient 1. Nevertheless, maximal pleural temperature was

Table 4. Mean values of hematological parameters

	WBC × 10^3	RBC × 10^6	Hemoglobinemia g/dl	Hematocrit %	Platelets × 10^3	TP %
J–1	6.7	4.8	13.7	41.4	328	97
et	0.8	0.5	1.2	2.8	42.8	5
T 0	5.9	3.3	12.6	38.8	260.2	92.6
et	3	1.7	1.8	5.5	65	6.3
T 30	12.9	4.1	11.2	34.8	219	90.8
et	1.6	0.5	2.3	6.1	48.9	13.4
T 60	11.6	4	11.8	34.6	262.8	85.8
et	2.2	0.4	1.1	2.9	77.7	10.8
T 90	14.8	4.2	12.7	37.1	239.5	81.3
et	4.4	0.3	1.7	4.8	34.3	4.3
T 360	14.6	3.8	11.4	33.5	286.3	87
et	2.6	0.3	1.5	4	95.8	4.6
T 480	13.6	3.7	11	33.3	299.5	83.5
et	2	0.4	1.1	4.7	104.3	8.9
J 1	10.8	3.5	9.9	30.7	263.2	77
et	3	0.3	1	4.2	104.3	9.4

TP = Percentage of prothrombin; *et* = ecartype. The mean values of the different parameters are measured during preoperative day J–1, at the beginning of the device T 0, after 30 min T 30, 60 min T 60, 90 min T 90, 360 min T 360, 480 min T 480 and after 1 day J1.

42.6 °C and maximal pulmonary temperature never exceeded 37.4 °C.

The blood levels of mitomycin C (table 5) never reached the systemic toxic level (1 mg/ml).

Discussion

Malignant pleural disease is considered to be unresectable. Repeated thoracocentesis and/or chest tube drainage with drug instillation is a common method used to alleviate the symptoms. However, the average survival period of patients with pleural metastasis who received such conservative therapy was 6 months or less [6]. The prognosis of pleural mesothelioma is also unfavorable with or without treatment, the median survival reported in the literature varies from 7 to 16 months [1, 3]. Surgery seems to prolong survival [2, 7]. Complete remission of a pleural mesothelioma has been described in one case with cisplatyl [8]. Combination chemotherapy (cyclophosphamide, cisplatyl, mitomycin [9, 10] can be used. Boutin et al. [4] described the use of recombinant γ-interferon by the intrapleural route as treatment of malignant pleural effusion. Moore et al. [5] described complete resolution of pleural effusion in 3 dogs treated with intrapleural cisplatin. In spite of these different treatments, the prognosis remains unfavorable [11]. It may be that a combination of these different treatments could increase the survival rate [12].

Some authors [13–15] proposed hyperthermia via ultrasound or radiofrequency for lung cancer with carcinomatosis, hyperthermia seems to be effective against malignant cells. The efficacy of hyperthermia against tumors has been demonstrated in vitro [16] and in vivo [17–19]. On the other hand, hyperthermia has the potential to modify the cytotoxicity of anticancer drugs and to convert some other drugs to cytotoxic agents having no considerable toxicity at normal temperature [20, 21]. Mitomycin cytotoxicity, as for cisplatin, was found to be enhanced by heating at 41–42 °C for 1 h [6, 16].

Our experience of intraperitoneal chemohyperthermia IPCH [18, 19] shows that the device is safe and reliable for patients with peritoneal carcinomatosis so we tried this method in the pleural cavity. The methodology and

Table 5. MMC levels for each patient treated by ITCH

Patient	T 30	T 60	T 90	T 360	T 480	J 1
1	0.06	0.08	0.08	<0.005	0.01	0.01
2	0.005	0.005	0.005	0.08	<0.005	<0.005
3	0.1	0.03	<0.005	<0.005	<0.005	
4	0.07	0.07	0.07	<0.005	<0.005	<0.005

Mitomycin C in mg/l. The data are measured during ITCH, after 30 min T 30, 60 min T 60, 90 min T 90, 360 min T 360, 480 min T 488, and after 1 day J1.

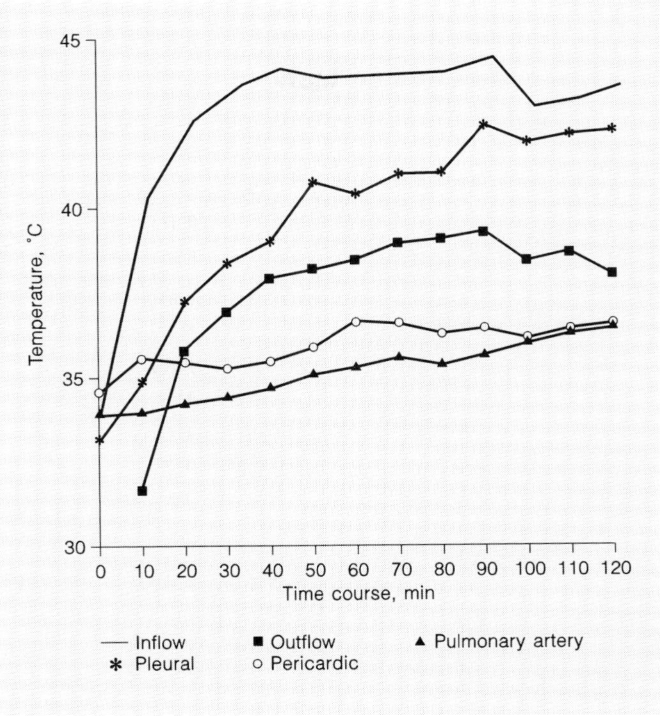

Fig. 1. Temperatures.

material used for IPCH are the same. The differences are the duration of the protocol and the temperature. The duration of hyperthermia is lower than the intraperitoneal method, 60 instead of 90 min, as the adequate temperature is reached more rapidly. To prevent the pulmonary effects of high temperature (edema), the inflow temperature never exceeded 45 °C, and the pleural temperature 43 °C [22]. As with IPCH, no mortality or morbidity were observed during ITCH, whereas biological variations were transitory. The blood levels of chemotherapy were lower than during IPCH [18]. Two hypotheses could explain these differences: (1) lower temperature in the pleural than in the peritoneal cavity; (2) lower absorption capacity of the pleural membrane. This lower absorption, and the predominant local effect could also explain the good tolerance of this treatment as noted by Bogliolo et al. [23].

Finally, intrapleural chemotherapy associated with intrapleural hyperthermia is a safe and reliable therapeutic device. Obviously, it is too early to get any valid conclusions on this new therapeutic device. Further studies could be of interest to evaluate its antitumoral efficacy.

References

1 Branscheid D, Krysa S, Bauer E, Bulzebruck H, Schirren J: Diagnostic and therapeutic strategy in malignant pleural mesothelioma. Eur J Cardiothorac Surg 1991;5:466–473.

2 Rusch VW: Diagnosis and treatment of pleural mesothelioma. Semin Surg Oncol 1990;6:279–285.

3 Ruffie P: Traitements des tumeurs pleurales. Rev Prat 1990;40:1951–1953.

4 Boutin C, Viallat JR, Van Zandwijk N: Activity of intrapleural recombinant gamma-interferon in malignant mesothelioma. Cancer 1991; 67:2033–2037.

5 Moore AS, Kirk C, Cardona AAD: Intracavitary cisplatin chemotherapy experience with six dogs. J Vet Intern Med 1991;5:227–231.

6 Kodama K, Doi O, Tatsuda M, Kuriyama K, Tateishi R: Development of postoperative intrathoracic chemothermotherapy for lung cancer with objective of improving local cure. Cancer 1989;64:1422–1428.

7 Rusch VW, Piantadosi S, Homes EC: The role of extrapleural pneumectomy in malignant pleural mesothelioma. J Thorac Cardiovasc Surg 1991;102:1–9.

8 Menard O, Seigneur J, Lamy P: Mise en remission d'un mésothéliome pleural malin par du cisplatine à forte dose en monochimiothérapie. Rev Mal Respir 1991;8:413–414.

9 Unasawasdi T, Dhingra HM, Charnsangavej C, Luna A: A case report of malignant pleural mesothelioma with long-term disease control after chemotherapy. Cancer 1991;67:48–54.

10 Ardizzoni A, Rosso R, Salvati F, Fusco V, Cinqugrana A, De Palma M, Serrano R, Pennuci MC, Soresi E, Crippa M: Activity of doxorubicin and cisplatin combination chemotherapy in patients with diffuse malignant pleural mesothelioma. Cancer 1991;67:2984–2971.

11 Ball DL, Cruickshank DG: The treatment of malignant mesothelioma of pleura: Review of a 5-year experience. Am J Clin Oncol 1990;13: 4–9.

12 Sugarbaker DJ, Heher EC, Lee TH, Couper G, Mentzer S, Corson JM, Collins JJ, Shemin R, Pugatch R, Weissman L: Extrapleural pneumectomy, chemotherapy and radiotherapy in the treatment of diffusive malignant pleural mesothelioma. J Thorac Cardiovasc Surg 1991; 102:14–15.
13 Sekins S, Keilman G, Anderson T: Lung cancer hyperthermiavia ultrasound and pfc liquids. Vth International on Clinical Hyperthermia, Kyoto, 1988, p 6-c1.
14 Karasawa K, Muta N, Takahashi N, Aoki Y, Nakagawa K, Sakata K, Iio M: RF capacitance hyperthermia and radiolotherapy of lung cancer. Vth International on Clinical Hyperthermia, Kyoto, 1988, p 6-c3.
15 Terashima H, Nakata H, Maeoka N, Kunugida N, Tsuchiya T, Yoshimtasu H: Local Rf-wave hyperthermia (8 MHz) combined with radiotherapy in advanced lung cancer. Vth International on Clinical Hyperthermia, Kyoto, 1988, p 6-c3.
16 Barlogie B, Corry M, Drewinko B: In vitro thermochemotherapy for human colon cancer cells with cisdichloro-diaminoplatinium. Cancer Res 1980;40:1165–1168.
17 Fujimoto S, Shrestha RD, Kokubun M, Ohta M, Takahashi M, Kiuchi S, Okui K, Miyoshi T, Arimizu N, Takazawa H: Intraperitoneal hyperthermic perfusion combined with surgery effective for gastric cancer patients with peritoneal seedings. Ann Surg 1988;208:36–41.
18 Gilly FN, Carry PY, Sayag AC, Braillon G, Panteix G, James I, Chatelard Ph, Manchon M, Degouttes S: Intraperitoneal chemohyperthermia (IPCH) with mitomycin C (MMC) for far advanced cancer with peritoneal seedings: Our first experience. Jpn J Hypertherm Oncol 1991;7:208–218.
19 Gilly FN, Carry PY, Sayag AC, Panteix G, Manchon M, Rochette A, Peix Jl, Baulieux J, James I, Braillon G: Tolerance of intraperitoneal chemohyperthermia (IPCH) with mitomycin C (MMC): In vivo study in dogs. Int J Hyperthermia 1992;8:659–666.
20 Turano C, Ferraro A, Strom R: The biochemical mechanism of selective heat sensitivity of cancer cells. Eur J Cancer 1970;6:67–72.
21 Hahn GM: Potential fir therapy of drugs and hyperthermia. Cancer Res 1979;78:883–892.
22 Mustafa KY, Selig WM, Burhop KE, Minnear FL, Malik AB: Hyperthermia-induced pulmonary edema. J Appl Physiol 1986;60:1980–1985.
23 Bogliolo GV, Lerza R, Bottino GB, Mencoboni MP, Pannacciulli IM, Vannozzi M, Fulco RA, Merlo F, Esposito M: Regional pharmacokinetic selectivity of intrapleural cisplatin. Eur J Cancer 1991;27:839–843.

Oncology 1993;50:353–361

Homayoon Shidnia
Ned Hornback
Gwen Ford
Rong-Nian Shen

Indiana University Medical Center
Indianapolis, Ind., USA

Clinical Experience with Hyperthermia in Conjunction with Radiation Therapy

Key Words
Radiation therapy
Hyperthermia
Retreatment of irradiated tissue

Abstract
The authors have reviewed the medical records of 421 sessions of hyperthermia treatments in 73 patients treated between 1987 and 1992 at the University Heights Cancer Center and the Indiana University Medical Center in Indianapolis, Ind. Temperatures attained during the course of therapy on each patient have been averaged and the results were evaluated for complete, partial, or no response. All patients had previously failed conventional radiation therapy, chemotherapy and surgery. Responses were defined as: (1) Complete response – lesions that completely disappeared during treatment and the response was sustained for a minimum of 6 months. (2) Partial response – lesions that underwent a reduction in size of greater than 50%. (3) No response – less than 50% reduction in tumor size during the course of treatment. Response varied somewhat according to histology and anatomical site of treatment; however, complete response was achieved in 45%, partial response in 48% (for an overall response of 93%), and no response in 7% of the patients. The response achieved varied with temperature attained and a minimum temperature of 40°C for 40 min produced the greatest number of responses. Response to hyperthermia was directly related to the temperature achieved and the length of time the temperature was applied.

Introduction

'In our experience, radiation therapy for post-radiation recurrence following a full dosage is contraindicated, since it almost invariably causes necrosis and fails to control the disease' [1].

'High-dose' radiation therapy up to the tolerance of normal tissue (i.e. 7,000 cGy), will produce excellent local tumor control. However, if the tumor recurs after high-dose radiation, the door is frequently closed for further irradiation of normal tissue, as it is very difficult to give additional treatment to the area without having major morbidity and complications. The use of hyperthermia alone or in conjunction with radiation therapy or chemotherapy offers an alternative to the 'wait, watch and send the patient home to die' policy. Encouraging results utilizing hyperthermia in previously irradiated tissues have been reported by several authors [2–8] and have opened the way for retreatment of recurrent tumors lying in previously heavily irradiated normal tissue.

The following events have been demonstrated to occur when hyperthermia is used in the treatment of malignant

Homayoon Shidnia, MD
Indiana University Medical Center
Indianapolis, IA 46202-5289 (USA)

0030–2414/93/0505–0353
$2.75/0

tumors [4]: (1) increased permeability of the cell membrane; (2) alteration of nucleic acid and protein; (3) irreversible damage to the cellular respiration; (4) release of lysosomes; (5) decrease in intracellular pH.

In general, the effects of hyperthermia are related to the overall temperatures achieved and can be summarized into the three following categories: (1) thermal damage due to the high temperature (the higher the temperature used in local hyperthermia, the greater the damage); (2) thermal enhancement of tumor kill when added to chemotherapy and radiation therapy (moderate temperature between 40 and 43 °C used in area or regional hyperthermia); (3) thermal immune stimulant seen at a lower temperature, i.e. 39–40 °C (seen in systemic or whole body hyperthermia).

Material and Methods

From 1987 until 1992, 421 sessions of hyperthermia treatment following radiation therapy were reviewed in 73 patients. In all patients conventional radiation therapy, chemotherapy and/or surgery had previously failed. All patients' tumors were confirmed histologically and some patients had more than one anatomical site of recurrence. Radiation therapy treatments were uniform in that patients with cutaneous and mucosal lesions received 5,000 cGy in 5 weeks except for malignant melanoma lesions which received a total dose of 3,500 cGy given in five weekly fractions of 700 cGy each. Radiation was given either with a variable electron beam or a 10-MV photon beam from the linear accelerator. Hyperthermia was given within half an hour following radiation therapy. The treatment plan was for the patient to receive 1 h of hyperthermia not to exceed 43 °C. Hyperthermia was given utilizing the 915-MHz Cheung Laboratories System 100. Thermometry was performed utilizing 6–8 thermal probes (Luxtron). For larger tumors, a minimum of one intratumoral probe was inserted into the tumor (probe 1) with an additional 5–7 surface skin probes. For superficial lesions, one temperature probe (probe 1) was placed on the skin beneath a 1 cm diameter and 1 cm thickness of tissue equivalent Superflap (Mick Radio-Nuclear Instruments) (to prevent cooling of the probe and to simulate subcutaneous temperature). Hyperthermia treatments were given twice per week for a total of eight treatments with the exception of malignant melanoma, which received a single hyperthermia treatment per week. The skin under the hyperthermia treatment port was air-cooled by fan. Despite air cooling, some patients were unable to tolerate 1 h of hyperthermia and some patients elected not to complete the scheduled eight treatments. Even though the net power ranged from 176 to 200 W/cm^2 for a maximum of 1 h, some lesions did not reach the planned tumor temperature of 43 °C.

A diagram of the tumor and placement of four probes was made and clinical photos of all the lesions were taken prior to treatment (fig. 1–16).

Thermal doses were collated and averaged based on the primary temperature tumor probe (probe 1), which in all patients was either intratumor or under the tissue equivalent 'superflab' in cases of superficial lesions. The results were then tabulated and compared with the responses achieved.

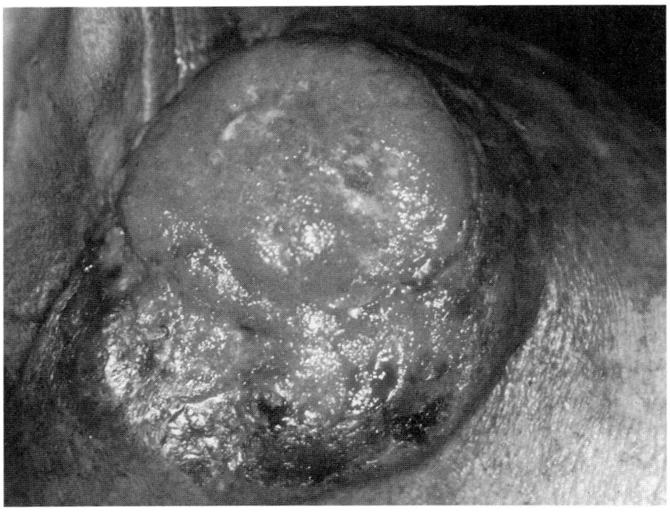
1

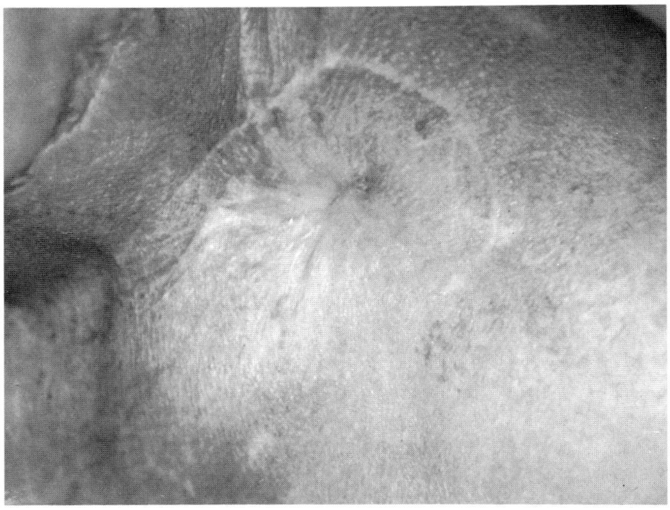
2

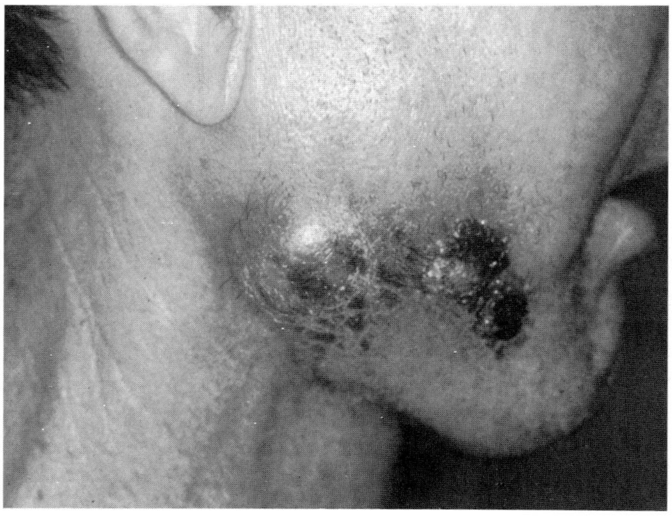
3

Fig. 1. Recurrent squamous cell carcinoma of FOM in left SCF after 5,000 cGy.

Fig. 2. Three months after 5,000 cGy and hyperthermia.

Fig. 3. SCC of right lower gum with bone and skin invasion.

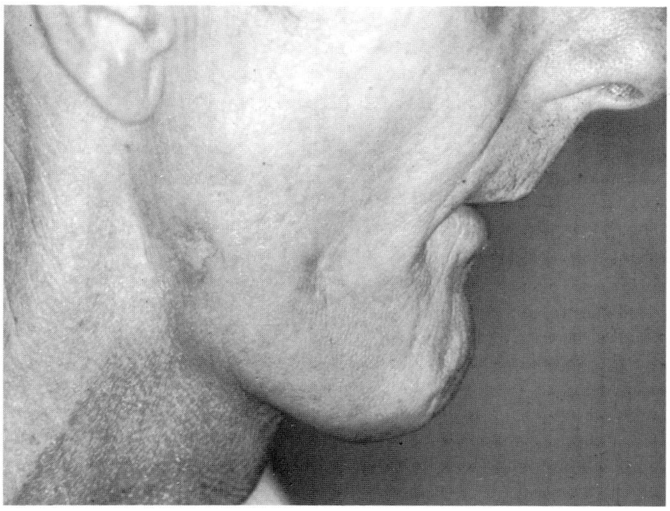

4

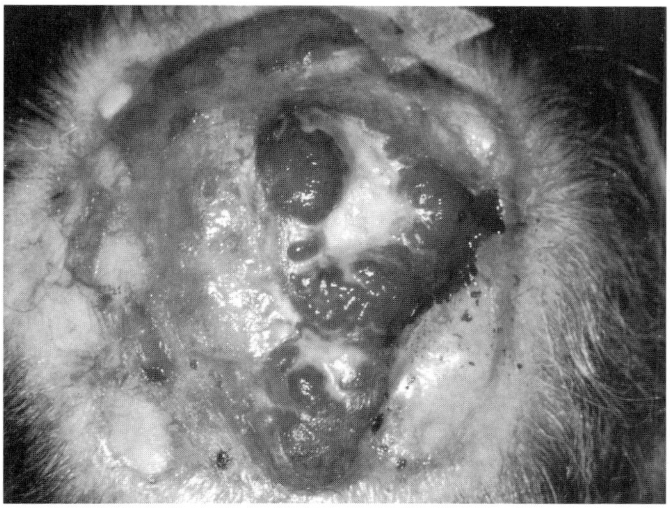

5

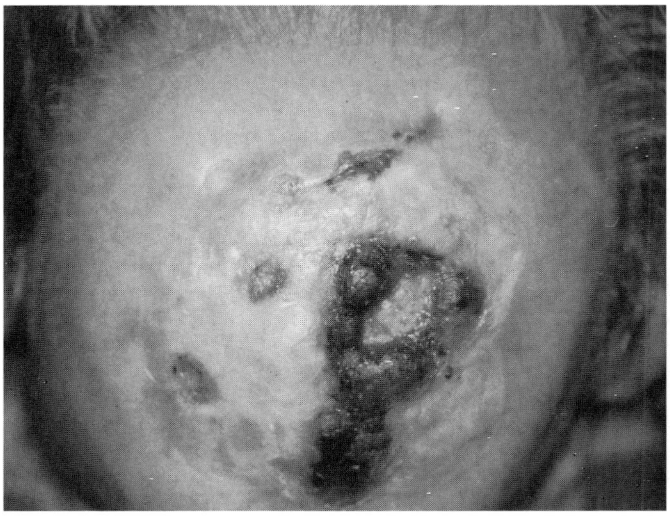

6

Fig. 4. Three months after 5,000 cGy and hyperthermia.
Fig. 5. Malignant melanoma of scalp.
Fig. 6. One month after 3,200 cGy and hyperthermia.

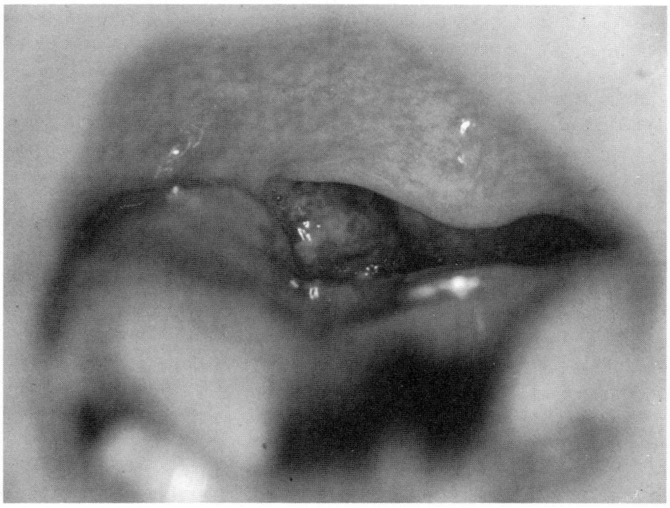

7

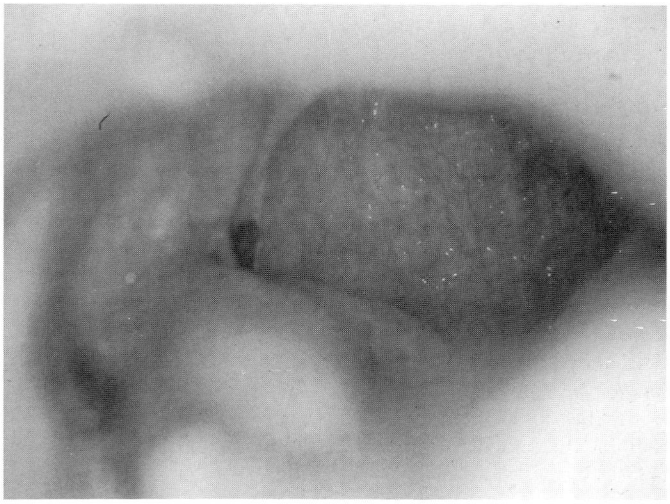

8

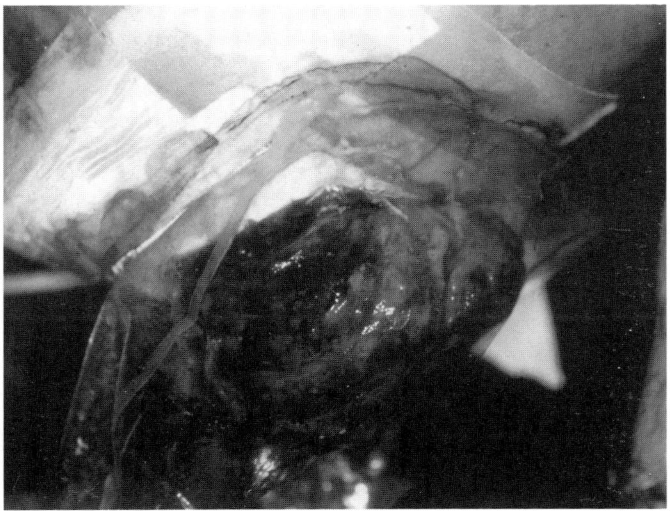

9

Fig. 7. Malignant melanoma of right tonsil.
Fig. 8. Three months after 3,200 cGy and hyperthermia.
Fig. 9. Malignant fibrohystiocytoma.

10
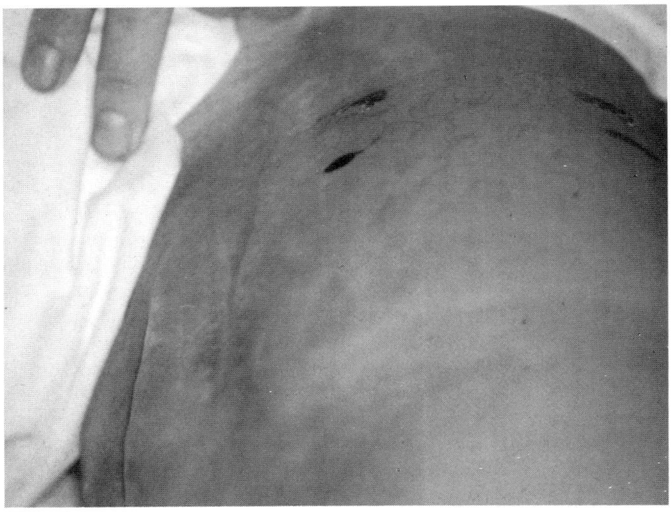

11
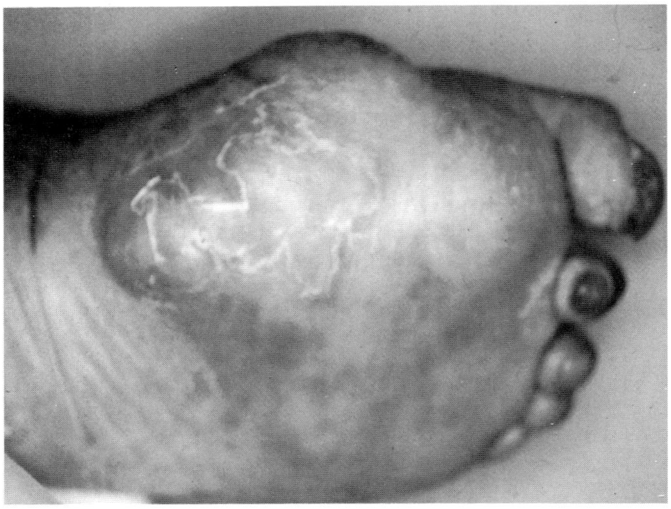

12
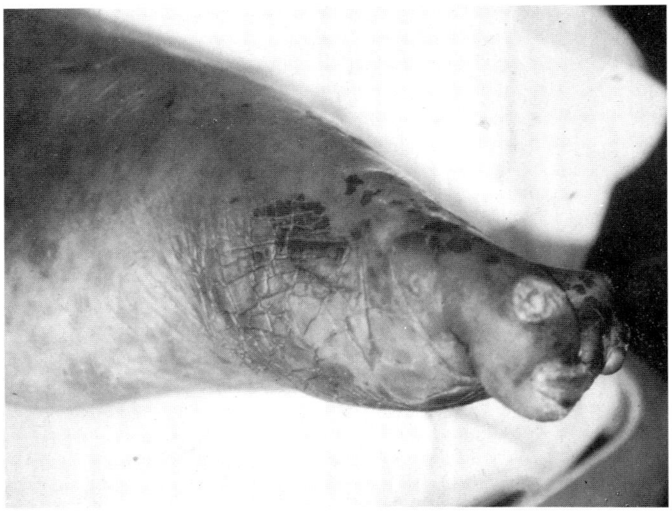

Fig. 10. Three months after 5,000 cGy and hyperthermia.
Fig. 11. Chondrosarcoma of left foot.
Fig. 12. Three months after 3,200 cGy and hyperthermia.

Table 1. Results of treatment with hyperthermia and radiation therapy (1987–1992)

Anatomical site	Number of sites	Responses		
		CR	>50%	<50%
Skin, malignant melanoma	195	150 (77%)	39 (20%)	6 (3%)
Head and neck cancer	172	75 (44%)	89 (51%)	8 (5%)
Breast	29	8 (28%)	16 (55%)	5 (17%)
Soft tissue sarcoma	18	14 (78%)	3 (16%)	1 (6%)
Lung	4	1	2	1
GI	1		1	
GYN	1			1
Unknown primary	1		1	
Total	421	248	151	22

Table 2. Result of treatment of 369 sites in 66 patients with average of 5.6 treatments per patient for 1 h per Rx

Response	Patients	
	n	%
CR	24	36
>50	33	50
<50	9	14
Total	66	100

Results

Table 1 demonstrates the overall treatment responses in patients receiving hyperthermia and radiation. Complete response for malignant melanoma lesions was 77%. The complete response rate for head and neck tumors was 44%, breast tumors 28%, and soft tissue sarcomas 78%. The overall response rate (complete plus partial response) in melanoma was 97%, head and neck 95%, breast 83%, and soft tissue sarcoma 94%. For all tumors regardless of

13

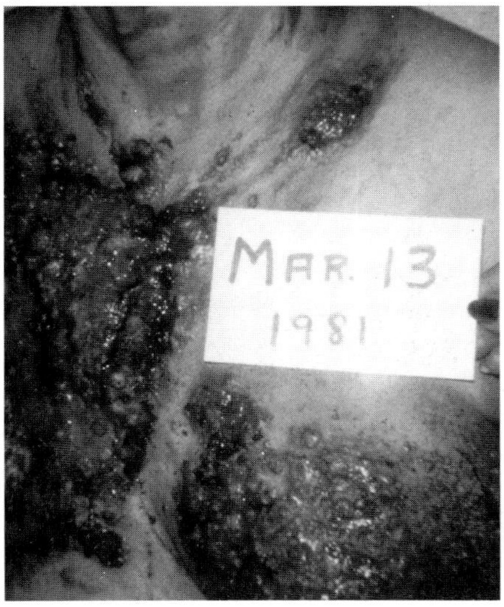

14

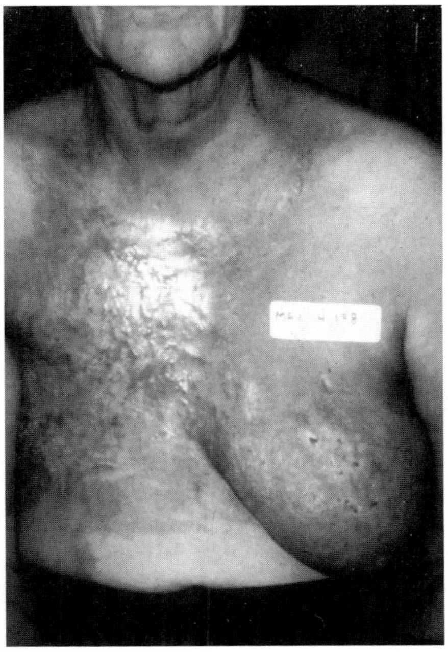

15

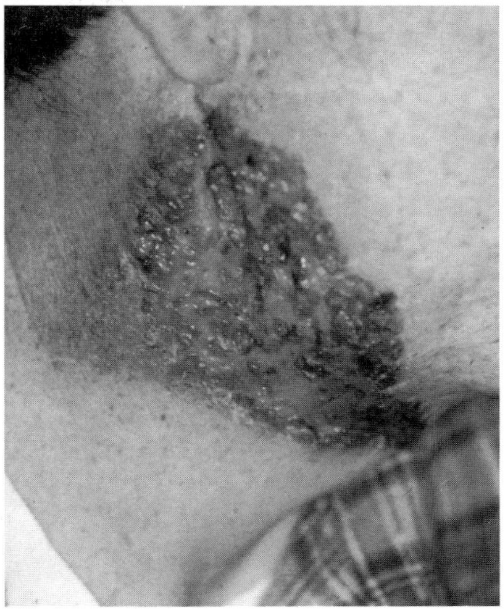

Fig. 13. Recurrent carcinoma of the breast after 6,500 cGy.

Fig. 14. One month after radiation and hyperthermia.

Fig. 15. Malignant melanoma with metastasis to neck node treated with 3,200 cGy, no hyperthermia on the right side.

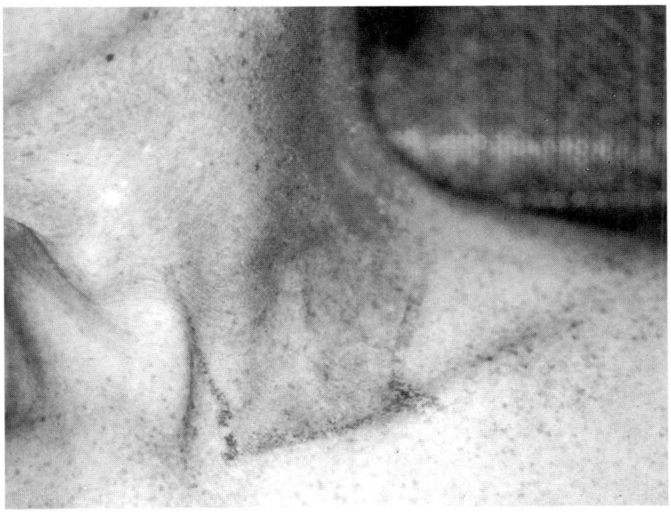

Fig. 16. Same patient as figure 15 treated with 3,200 cGy on the left side with hyperthermia.

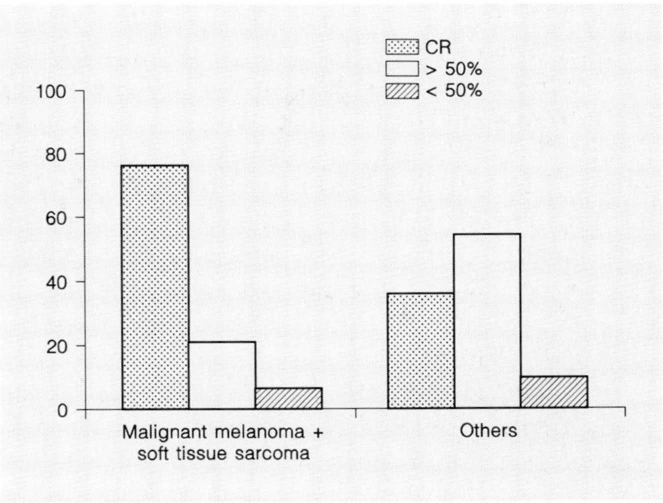

Fig. 17. Response to radiation and hyperthermia in all patients.

site, the overall response rate (complete and partial) was 94% (fig. 17).

Table 2 lists the number of patient response rates for patients who had a minimum of 6 months' follow-up. Three-hundred and sixty-nine lesions in 66 patients were evaluated in this study. Fifty-two lesions in 7 patients were excluded because of the lack of adequate thermometry measurements and the follow-up time being less than 6 months. As shown in table 2 in 66 patients, regardless of the number of lesions treated, the complete response was

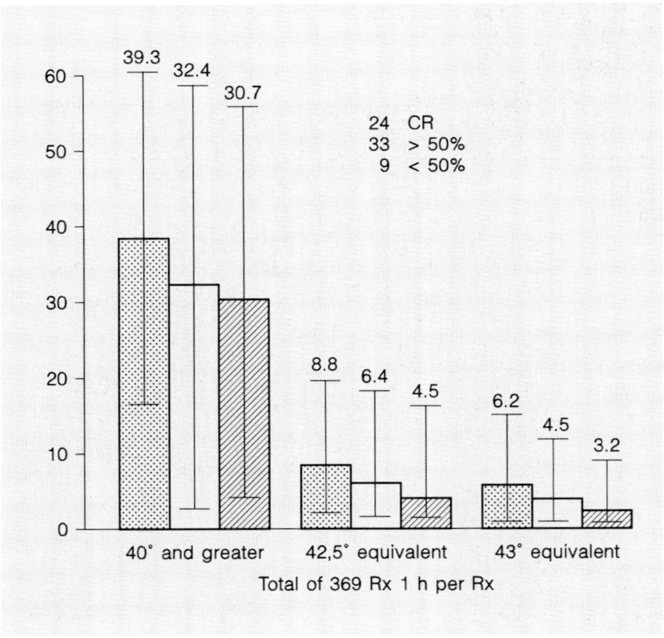

18

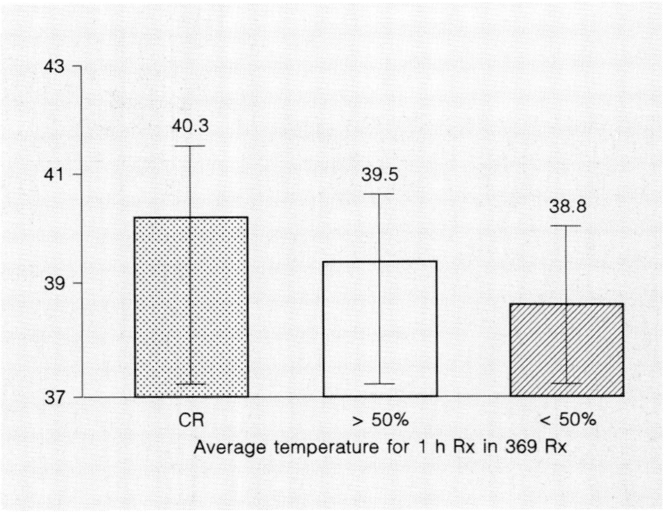

19

Fig. 18. Thermotherapy response according to different thermal ranges.
Fig. 19. Response with regard to average temperature.

36% and the overall response rate (complete plus partial response) was 86%. Fourteen percent were considered nonresponders or had a response rate of less than 50%.

Table 3 shows response rates of 369 lesions treated in the 66 patients. The complete response in all lesions was 45% with an overall response rate of 93%.

An attempt was made to correlate the relationship between thermal tumor dose achieved and the response obtained. Radiation therapy doses were uniform and hyperthermia treatments were kept constant. Patients were scheduled for eight hyperthermia treatments for 1 h twice a week. Some of the patients did not complete their full course of hyperthermia.

The tumor temperature recorded on a single probe (probe 1) during the course of hyperthermia was averaged in each patient. This temperature represented the highest temperature recorded in almost all of the patients. The averaged temperature was then used to calculate an equivalent time at temperatures of 40, 42.5 and 43°C using the Arrhenius plot ($R = e^{-\Delta H/[2T(T+1)]}$). The results are seen in figure 18 [9].

It was also noted that a minimum temperature of 40°C for a period of at least 40 min was required to produce a significant number of responses. This would be the equivalent of 10 min at 43°C according to the Arrhenius plot (fig. 19).

Table 3. Results of treatment of 369 sites in 66 patients with average of 5.6 treatments per patient for 1 h per Rx

Response	Sites	
	n	%
CR	164	45
>50	178	48
<50	27	7
Total	369	100

Discussion

The results of this study reveal an overall response rate in 66 patients of 93% for 369 sites (complete and partial responses). No response (response less than 50%) was seen in only 7%.

In his studies, Arcangeli [3] used higher temperatures of 42.5–43.5°C for 1 h and the overall response rate did not seem greater than what we have achieved at the lower temperatures. Arcangeli [3] showed that patients who received hyperthermia twice a week did better in terms of long-term control than patients who did not receive hyperthermia, and patients who received one treatment per week did not do significantly better than patients who

Table 4. Clinical studies directed to test the influence of the number of heat treatments (HT) on outcome

Author	Number of patients or lesions	°C/min	Number of HT per week	Total HT	% complete response	Statistical difference
Alexander et al.	48	42.5–44/60	1	4	42	NA
		42.5–44/60	2	8	21	
Arcangeli et al.	17	–	0	0	35[a]	–
	14	43.5/45	1	5	64[b]	b vs. a NS
	9	43.5/45	2	10	78[c]	c vs. a $p<0.05$
Kapp et al.	126	43–45/45	1	2	65	
		43–45/45	2	6	73	NS
Leopold et al.	8	42/60	1	5	37.5	$p=0.009$
	9	42/60	2	10	100	
Bicher et al.	78		5	25	55+38	partial response
	78		2	10	41+37	
Oleson et al.	108	40/60	1	?	44	NS
		40/60	2	?	44	
Valdagni et al.	7	42.5/30	2	2	86	NS
	10	42.5/30	2	6	80	

Table 5. Interstitial hyperthermia and radiation

Authors	Evaluable tumors n	Complete response %	Partial response %	No response %
Vora et al.	15	73	7	20
Oleson et al.	52	39	42	19
Cosset et al.	23	83	17	0
Puthawala et al.	43	74	26	0
Emami et al.	44	59	27	14
Lam et al.	31	61	36	3
French multicenter study (MINERVE)	96	61	33	6

received no hyperthermia. Bicher [8] reported that daily hyperthermia five times a week for a total of 25 times produced superior results over those which could be obtained with ten hyperthermia treatments twice a week (table 4).

In reviewing the literature [3, 8], patients who received 1–8 deep heat treatments did no better than patients who received 9–15 treatments. However, when patients received 25 treatments those patients were felt to have a better response. There is still a question concerning the most effective number of hyperthermia treatments that should be given per week. The number appears to be more than one; however, a higher number of hyperthermia treatments per week does not seem to increase local control (tables 5, 6).

Overgaard [2] has shown that the TER (thermal enhancement ratio) ranging from 1.5 to 4 could be expected. (tables 7, 8). The problem of thermal tolerance of tumor cells has been definitely established in laboratory studies [10], but clinically, we have not seen evidence of thermal tolerance.

Uniform temperatures throughout the tumor are virtually impossible to attain with existing hyperthermia equipment and it is often very difficult to reach the desired temperature. Failure to achieve the desired temperature should not discourage the physician from use of this very valuable palliative treatment modality as hyperthermia appears to enhance radiation effects even at the lower temperatures.

Table 6. Responses of superficial lesions to irradiation and irradiation plus hyperthermia

Institution	Evaluable patients (treatment trials)	Same patient comparisons	Irradiation alone %	Irradiation + hyperthermia %
Duke University	7	7	14	86
Memorial Hospital	86	59	33	80
M.D. Anderson	18	13	0	62
IMRS (Rome) Study				
I	26	26	42	73
II	16	16	37.5	67, 77[1]
III	15	15	33	87
Roswell Park	31	31	39	87
Stanford University	15	15	7	47
Malmö[2]	73/33	53/17	30	54
Tianjin[2]	124	62	29	68

Modified from Meyer JL: Cancer Res 1984;44(suppl 47):4745s.

1 Hyperthermia immediately after radiotherapy (77%) or delayed 4 h (67%).

2 Comprable lesions not necessarily in same patient.

Table 7. Malignant melanoma: Isodose TER

Authors	Number of tumors	Radiation alone, %	Radiation and heat, %	Isodose TER
Overgaard et al.	63	57	90	1.58
Gonzalez et al.	24	17	83	4.88
Kim et al.	149	43	67	1.56
Arcangeli et al.	38	53	76	1.43
Valdagni et al.	35	20	80	4.00
Emami et al.	116	24	59	2.45
Shidnia et al.	185	33	64	1.94

Note: Tumor response after treatment with the same radiation dose given either alone or combined with hyperthermia.

Table 8. Breast-chest wall recurrence: isodose TER

Authors	Number of tumors	Radiation alone, %	Radiation and heat, %	Isodose TER
Perez et al.	35	43	86	2.00
van der Zee et al.	40	0	24	$\gg 1$
Scott et al.	34	47	94	2.00
Gonzalez et al.	18	33	78	2.36
Lindholm et al.	66	35	70	2.00
Li et al.	42	36	73	2.03
Dunlop et al.	32	67	82	1.22
Overgaard et al.	14	40	78	1.94

Note: Tumor response after treatment with the same radiation dose given either alone or combined with hyperthermia.

Summary

Hyperthermia is a very effective and valuable tool which produces a high degree of response in a group of seriously ill patients who have little chance of responding to standard radiation therapy treatment. Thermometry is essential on every patient undergoing hyperthermia and is required for comparison of different heating techniques. Hyperthermia of 40–43°C for 1–2 h should be considered the goal for patients undergoing hyperthermia and radiation. The number of hyperthermia fractionations per week is unknown, but it appears that two treatments per week is sufficient to produce the enhancement of radiation. Additional hyperthermia treatments beyond two treatments per week does not appear to increase local control. A minimum follow-up of 6 months is required to evaluate length of tumor response.

References

1 Fletcher G: Textbook of Radiotherapy, ed 2. London, Henry Kimpton, 1973, p 159.
2 Overgaard J: The current and potential role of hyperthermia in radiotherapy. Int J Radiat Oncol Biol 1989;16:535–549.
3 Arcangeli G: One versus four heat treatments in combination with radiotherapy in metastatic mammary carcinoma. Int J Radiat Oncol Biol 1991;21:1569–1574.
4 Perez C, Brady L: Principles and Practice of Radiation Oncology, ed 2. Philadelphia, Lippincott, 1992, pp 396–446.
5 Hornback N: Advanced stage IIIB cancer of the cervix treatment by hyperthermia and radiation. Gynecol Oncol 1986;23:160–167.
6 Oleson J: Result of a phase II trial of hyperthermia and radiation therapy. 38th Ann Meet Abstr Radiation Res Soc, 1990.
7 Kapp D: Hyperthermia as an adjuvant to radiation in treatment of superficial metastasis. 34th Ann Meet Abstr Radiation Res Soc, 1986.
8 Bicher H: Direct relation between thermal dose and tumor response in hyperthermia therapy. Abstract, Ann Meet Int Clin Hyperthermia Soc, Dubna, 1991, p 8.
9 Sapareto S: Thermal dose determination in cancer therapy. Int J Radiat Oncol Biol 1984; 10:787–800.
10 Li G: Induced thermal tolerance and heat shock synthesis. Br J Cancer 1982;45:132.

Oncology 1993;50:362–365

C. Berny[a], A. Mialon[a]
M. Manchon[a], K.E. Le[a]
G. Panteix[a], P. Baltassat[a]
F.N. Gilly[b], P.Y. Carry[c]
A. Sayag[c], G. Braillon[b]

Departments of
[a] Biochemistry,
[b] Surgery and
[c] Anesthesia and Intensive Care,
Centre Hospitalier Lyon-Sud,
Pierre-Bénite, France

Biochemical Modifications in the Blood and the Heated Fluids during Intraperitoneal Chemohyperthermia

Key Words
Biochemical modification
Hyperthermia
Intraperitoneal perfusion
Tumor cell lysis

Abstract
The biochemical changes in blood during intraperitoneal chemohyperthermia (IPCH) were examined by carrying out complete assessments before and after the operation. These assessments were made up of 23 parameters: Na, K, Cl, CO_2, urea, creatinine, proteins, glucose, calcium, phosphates, magnesium, bilirubin, uric acid, lactic acid, CRP, ASAT, ALAT, CK, LDH, γ-GT, ALP, lipase, and amylase. Only 5 of these parameters showed significant changes: proteins, urea, ALP, γ-GT, lactic acid. The protein and urea levels decreased due to hemodilution induced by the perfusion of fluids. ALP and γ-GT levels decreased, possibly due to localized inhibition of secretion. Lactic acid levels increased due to the movement of lactates from the heated fluid into the blood. The study of biochemical changes within the heated fluids was made using the following parameters: CA 125, CA 19-9, CEA, ASAT, ALAT, CK, LDH, γ-GT, ALP, lipase, uric acid, phosphates, proteins, Na, K, Cl, urea, creatinine, and magnesium. Between the beginning and the end of IPCH, significant increases were found in the levels of CA 125 (+173%), proteins (+190%), ASAT (+130%), LDH (+103%), K^+ (+232%), PO_4 (+134%), and uric acid (+99%). These increases indicate the existence of a significant degree of cellular lysis.

Introduction

Since 1989 a pilot study has been conducted in Centre Hospitalier Lyon-Sud (CHLS) to evaluate intraperitoneal chemohyperthermia (IPCH) with mitomycin C (MMC) in the treatment of digestive cancers with peritoneal carcinomatosis [1]. IPCH is performed under general anesthesia and previously induced general hypothermia (32 °C). After surgical exeresis of the primary tumor, 6 heated liters (42–43 °C) of peritoneal dialysis fluid containing 10 mg/l mitomycin C are propelled inside the abdominal cavity of the patient using a peristaltic pump (200 ml/min) in a sterile closed circuit during a period of 90 min. This technique was first used by Spratt et al. [2], Koga et al. [3] and Fujimoto et al. [4].

Initially, at the request of the intensive care staff, we tried to discover whether, on a biological level, this new technique caused significant changes or whether, on the contrary, it was well tolerated.

C. Berny
Department of Biochemistry (3A)
Centre Hospitalier Lyon-Sud
F-69310 Pierre-Bénite (France)

0030–2414/93/0505–0362
$2.75/0

Secondly, we were interested in the circulating fluid in the peritoneum with a view to the possibility of identifying indicators of tumoral cell lysis therein.

Materials and Methods

The first objective was to study the biochemical changes in the blood induced by the hyperthermochemotherapy. 23 patients were included in this study. Primary tumors were gastric adenocarcinoma (n = 13), pancreatic adenocarcinoma (n = 3), peritoneal malignant mesothelioma (n = 2), ovarian carcinoma (n = 2), breast adenocarcinoma (n = 1), colon adenocarcinoma (n = 1). We carried out a complete biochemical blood assessment at time 0 and after IPCH. The following parameters were measured: Na, K, Cl, CO_2, urea, creatinine, proteins, glucose, calcium, phosphates, magnesium, bilirubin, uric acid, lactic acid, CRP, ASAT[1], ALAT[1], CK[1], LDH[1], γ-GT, ALP[1], amylase, and lipase. With the exception of lactic acid levels measured on ACA SX (Dupont, Wilmington, Mass., USA) and CRP levels measured on COBAS MIRA (Roche, Basel, Switzerland), all of these parameters were measured on a Synchron CX7 (Beckman, Brea, Calif., USA). All the blood estimations were carried out rapidly as soon as the samples arrived at the laboratory. Measurement of enzyme activities was carried out at 37 °C following the given methodology, where it exists.

The second objective was to study the indicators of tumoral cell lysis within the heated fluids. 14 patients, from whom samples of the heated fluid had been taken at time 0 and after IPCH, were included in this study. Primary tumors were gastric adenocarcinoma (n = 8), pancreatic adenocarcinoma (n = 3), colon adenocarcinoma (n = 1), peritoneal malignant mesothelioma (n = 1), ovarian carcinoma (n = 1). As possible molecular candidates for the title of 'indicator of cellular lysis' the following substances were used: (1) tumor markers: CA 125, CA 19-9 and CEA; measurements were carried out on an IMX (Abbott, Chicago, Ill., USA); (2) enzyme activities: ASAT, ALAT, CK, LDH, γ-GT, ALP, lipase; measurements were carried out on a Synchron CX7 as for the blood; (3) biochemical substrates: uric acid and phosphates, measured on a Synchron CX7 (Beckman) as for blood, and (4) total proteins, measured on an Astra 8 (Beckman).

We completed the assessment above with the inclusion of the traditional parameters, Na, K, Cl, urea, creatinine and magnesium. These last measurements were carried out on a Synchron CX7 (Beckman).

Finally we measured the hemoglobin of these fluids in order to be able to quantify, if necessary, the degree of contamination of the fluids by the patient's blood. This measurement was made using a spectrophotometric method on a Uvikon Roche (Kontron, St-Quentin-en-Yvelines, France).

All measurements were carried out on fluids which had been conserved at –20 °C and subsequently thawed.

Part of the discussion will be based on the composition of the 'Dianeal glucose' (Baxter) fluid, noted in table 1.

[1] Recommended by the SFBC.

Table 1. Composition of the Dianeal glucose Baxter fluid

Osmolarity, mosm/l	347
pH	5.5
Na, mmol/l	132
Ca, mmol/l	1.75
Mg, mmol/l	0.75
Cl, mmol/l	102
Lactates, mmol/l	35

Analysis of Results

For each parameter 'X' and for each medium (blood or heated fluid), we had a series of 'n' paired results, XT0 and XTFinal, corresponding to the number of patients included.

To compare the averages of these two paired series, the difference (d) between the 2 values (XTF – XT0) was calculated for each pair and then the average of the differences ($\bar{d}$) was compared to zero using the bilateral Student's t test, for which the formula is:

$$t = \frac{\bar{d}}{s/\sqrt{n}},$$

where s = standard deviation of the differences.

Results

Biochemical Changes in the Blood Caused by Hyperthermochemotherapy

The table 2 shows only the 5 parameters, of the 22 tested, for which significant differences were found between the values before and after hyperthermochemotherapy.

The other 18 parameters (Na, K, Cl, CO_2, creatinine, glucose, Ca, PO_4, Mg, bilirubin, uric acid, CRP, ASAT, ALAT, CK, LDH, amylase, lipase) do not show any significant variation, at 1% confidence limits, during the hyperthermochemotherapy.

Search for an Indicator of Tumor Cell Lysis in the Fluids

Conservation of the enzymes in a heated fluid was previously studied in vitro in the laboratory. Five samples of Dianeal glucose fluid were overloaded, using serum which had a very high level of enzyme activity. Each sample of overloaded fluid was then divided into 5 aliquots.

Aliquot 1 was kept at an ambient temperature; aliquots 2–5 were placed in a waterbath at 44 °C. Aliquot 2 was removed after 30 min, 3 after 60 min, 4 after 90 min and 5 after 120 min.

Table 2. Significant changes observed in the patients' blood during IPCH

Tests	Mean T0	sT0	Mean TF	sTF	Mean difference	t	% variation
Proteins, g/l	53.8	10.2	48.8	11.6	−4.333	−4.775	−9
Urea, mmol/l	3.57	1.63	2.89	1.33	−0.74	−4.430	−19
γ-GT, IU/l	37.7	32.7	29.6	30.4	−6.529	−2.921	−21
ALP, IU/l	75.6	40.3	62.3	45.5	−9.176	−3.653	−18
Lactic acid, mmol/l	2.2	1.06	3.6	1.42	1.417	5.500	+64

Table 3. Significant changes observed in the fluids during IPCH

Tests	Mean T0	sT0	Mean TF	sTF	Mean difference	t	% variation
CA 125, U/ml	12.28	11.34	33.57	15.7	20.57	4.84	+173
K, mmol/l	0.76	0.36	2.53	0.64	1.77	11.41	+232
Urea, mmol/l	0.73	0.43	2.17	0.70	1.44	6.86	+197
Creatinine, μmol/l	31.8	5.3	44.4	5.7	12.57	6.28	+40
PO_4, mmol/l	0.19	0.15	0.445	0.18	0.255	4.90	+134
Mg, mmol/l	0.70	0.03	0.58	0.04	−0.118	−7.46	−16
Uric acid, μmol/l	33.4	16	66.4	24	32.28	6.48	+99
ASAT, IU/l	5.5	2.5	12.7	8.03	7.21	4	+130
LDH, IU/l	74.2	98.8	151.2	1.08	76.85	5.22	+103
Proteins, g/l	1.23	1.30	3.57	1.40	2.34	7.8	+190

Enzyme activities of these aliquots were measured at the end of this experiment. All the enzyme activities remained thermostable for up to 120 min at 44 °C (deviation <5%).

Table 3 shows the results obtained in the patients' heated fluids. As in the previous section, table 3 shows only those parameters which had undergone a significant change in concentration within the fluid during the operation, at 1% confidence limits.

Discussion

Biochemical Changes in the Blood

There was no variation observed in 18 of the parameters measured during hyperthermochemotherapy. The question is how to explain the changes in the other 5 parameters.

For urea and protein levels, hemodilution linked with the perfusion employed (using Ringer's solution or serum glucose) could explain the decrease in these two parameters. These results concur with those of Fujimoto et al. [4, 5].

For ALP and γ-GT, localized inhibition of their secretion by the biliary tracts and/or the intestines could explain the reduction in those enzymes with a short half-life (1 h for intestinal ALP which represents 60% of the total ALP level).

For lactic acid, the most probable explanation of the observed increase of this parameter in the blood is the movement of lactates from the fluid (35 mmol/l) into the blood circulation.

It is normal to obtain high standard deviations for all parameters at T0 and TF since this reflects the diversity of results from one patient to another.

Modifications within the Fluids

Study of the averages obtained for each parameter in the fluids sampled at T0 shows that the fluids had not been properly collected at the very beginning of the session. Close inspection of the results shows, in fact, that the composition of 9 of the fluids at the beginning was almost the same as that of the initial fluid, the other samples having been collected later. Although this phenomenon would only tend to reduce the significance of our results we have nevertheless included these fluids in our study.

Two phenomena could explain the observations made on the fluids: (1) cellular lysis alone could explain the increase in CA 125, ASAT, LDH and in the proteins, and (2) the difference in concentration between the fluid and the circulating blood levels could explain the variations in urea, creatinine and magnesium due to 'dialysis':

The interaction of these two phenomena contributes to the increase in the fluid levels of potassium, phosphates and uric acid, but it is difficult to estimate the actual contribution of each meachanism for these parameters. The same applies for the standard deviations.

It is not possible to compare our results with those of other authors since the literature review carried out on this subject showed a paucity of references.

Conclusion

The results obtained from the blood samples show that hyperthermochemotherapy is a technique that was well tolerated biologically by the patients who were treated in this way. The results obtained from the fluids show that the increased concentrations observed in CA125 (+173%), proteins (+190%), ASAT (+130%) and LDH (+103%) indicate the existence of cellular lysis; the increased concentrations of K^+ (+232%), PO_4 (+134%) and uric acid (+99%) support this hypothesis. It would be interesting to compare, for each of our patients, the increase in these 7 parameters and the survival period of the patient, to see whether the two phenomena could be linked.

References

1 Gilly FN, Carry PY, Sayag A, Braillon G, Panteix G, James I, Cheatesard P, Manchon M, Desgoutte S: Intraperitoneal chemohyperthermia (IPCH) with mitomycin C (MMC) for far advanced cancer with peritoneal seedings: Our first experience. Jpn J Hyperthermic Oncol 1991;7:208–218.

2 Spratt JS, Adcock RA, Muskovin M: Clinical delivery system for intraperitoneal hyperthermic chemotherapy. Cancer Res 1980;40:256–260.

3 Koga S, Hamazoe R, Maeta M, Shimizu N, Mukarami, A, Wakatsuki T: Prophylactic therapy for peritoneal recurrence of gastric cancer by continuous hyperthermic peritoneal perfusion with mitomycin C. Cancer 1988;61:232–237.

4 Fujimoto S, Shresta RD, Kokubun M, Ohta M, Takahashi M, Kobayashi K, Kiuchi S, Okui K, Miyoshi T, Arimizu N, Takamizawa H: Intraperitoneal hyperthermic perfusion combined with surgery effective for gastric cancer patients with peritoneal seeding. Ann Surg 1988; 208:36–41.

5 Fujimoto S, Shresta RD, Kokubun M, Kobayashi K, Kiuchi S, Takahashi M, Konno C, Ohta M, Koike S, Kitsukawa Y, Mizutani M, Okui K: Clinical trial with surgery and intraperitoneal hyperthermic perfusion for peritoneal recurrence of gastrointestinal cancer. Cancer 1989; 64:154–160.

Oncology 1993;50:366–370

G. Panteix[a], M. Guillaumont[a]
L. Cherpin[a], J. Cuichard[a]
F.N. Gilly[b], P.Y. Carry[c]
A. Sayag[c], B. Salle[b]
A. Brachet[c], J. Bienvenu[a]
P. Baltassat[a], G. Braillon[b]

Departments of
[a] Biochemistry,
[b] Surgery and
[c] Anesthesia and Intensive Care,
Centre Hospitalier Lyon-Sud,
Pierre-Bénite, France

Study of the Pharmacokinetics of Mitomycin C in Humans during Intraperitoneal Chemohyperthermia with Special Mention of the Concentration in Local Tissues

Key Words
Chemohyperthermia, intraperitoneal
Mitomycin C
Pharmacokinetics

Abstract
The aim of this work is to estimate the 24-hour distribution and elimination of mitomycin C (MMC) during and after intraperitoneal chemohyperthermia (IPCH) in 18 patients (13 gastric adenocarcinoma, 3 pancreatic adenocarcinoma, 2 malignant mesothelioma) who received 60 mg MMC during 90–120 min in 6 liters of heating solution HS; (42 °C) or HS flowing at 0.4 liters/min in a closed circuit. MMC assay in the serum, urine, HS and in local biopsies were performed by high performance liquid chromatography. The amount of MMC in HS decreased by 54.1 ± 13.6% during IPCH. The maximum MMC levels in serum reached 0.4 ± 0.18 mg/l 45 min after the start of IPCH, then rapidly decreased. Only 1.77 ± 0.93 mg were recovered in urine in 24 h. These data are consistent with a large and rapid absorption, mostly in local tissue, demonstrated by the level in 7 post-IPCH biopsies (8.3 ± 7.6 mg/kg).

Introduction

The in vitro properties of heat on cancer cells have been known since 1970 [1]. The in vitro enhancement of the cytotoxicity of anticancerous drugs, such as mitomycin C (MMC), by the heat has been clearly demonstrated [2–4]. Similar reports have shown the toxic properties of in vivo hyperthermia [5, 6], and the in vivo enhancement of drug cytotoxicity by heat [7] has been described previously.

After first reports [8, 9], we began using a combined therapy (intraperitoneal drug and intraperitoneal hyperthermia) in human peritoneal carcinomatosis from digestive cancer. This combined therapy was called intraperitoneal chemohyperthermia (IPCH) with MMC and gave encouraging preliminary results [10].

The aim of the present study was to estimate the pharmacokinetics of MMC during and after IPCH and the eventual relationship between pharmacokinetic parameters and biological and clinical effects.

Dr. G. Panteix
Département de Biochimie (Bâtiment 2A)
Secteur Jules-Courmont
Centre Hospitalier Lyon-Sud
F–69310 Pierre-Bénite (France)

0030–2414/93/0505–0366
$2.75/0

Materials and Methods

Patients

From 1989 to 1991, 18 patients were included in the study: 9 men and 9 women (mean age 54.9 ± 10.3 years; mean weight 62.5 ± 10.7 kg). Primary tumors were gastric adenocarcinoma (n = 13), pancreatic adenocarcinoma (n = 3) and peritoneal malignant mesothelioma (n = 2).

Ten patients had large malignant ascites with a positive cytologic examination at the time of diagnosis. Surgical resection of the primary tumor was only possible for 7 patients, while 11 had unresectable primary tumors.

Inclusion Criteria

They were: (a) age <65 years; (b) cytologic and histologic confirmation of peritoneal carcinomatosis; (c) no extra abdominal metastases at the time of diagnosis; (d) acceptable cardiorespiratory function as checked by preoperative respiratory tests, ECG, and cardiac ultrasonography, and (e) informed consent of the patient and family.

IPCH Device

Under general anesthesia and mechanical ventilation, general 32 °C hypothermia was induced using ice bags, cold raps on the legs and a cooling hat. Surgical exploration of the abdominal cavity was performed, taking biopsies and eventual exeresis of the primary lesion. At the end of surgery, 2 silicone drainages (William Harvey, WH, Bard CP, USA) were inserted under the left and right diaphragmatic cupules and were called inflow drainage. A third one was inserted into the Douglas pouch and was called outflow drainage. thermic probes (Mallinckrodt, Medexel, France) were inserted into liver, parietal peritoneum and the Douglas pouch. The laparotomy was closed and the 3 WH drainages were connected to a closed sterile circuit in which 6 liters of peritoneal liquid were propelled using a roller pump (0.2 liter/min).

MMC was included in the closed sterile circuit at a rate of 10 mg/l, making a total dose of 60 mg/patient. This closed sterile circuit was heated through a thermic exchanger. IPCH was carried out during 90–120 min with an inflow temperature of 46–48 °C and an outflow temperature of 41–43 °C, and a local temperature of 42 °C.

For 7 patients, an immediate post-IPCH malignant node biopsy was surgically performed. During IPCH (0, 15, 30, 45, 60, 90, 120 min) and after IPCH (180, 240, 300, 360, 600, 1,440 min) blood, urine, perfusates, and then the peritoneal drain were sampled.

MMC Determination in Blood, Urine, Perfusates (or Heating Solution)

MMC concentrations were determined by high performance liquid chromatography (HPLC) in the various samples with a method adapted from Den Hartig et al. [11]. MMC was first extracted from the biological sample by chloroform:propanol 50:50 v:v (4 ml). After evaporation of the organic layer, dried residues were solubilized in 200 µl of the mobile phase. MMC chromatography used a Lichrosorb C18 5 µg, 150 × 4.6 mm ID (Merck, FRG) as a column and with acetonitril:water 25:75 v:v mixture as a mobile phase flowing at 1 ml/min.

MMC was detected by UV absorption at 360 nm. At each run, prior to MMC determination in samples from patients, serum spiked with known amounts of MMC were injected. The method was validated by using sera with a known amount of MMC.

The intradays and interdays reproducibilities were given in terms of the coefficient of variation (CV, %). The intraday reproducibility at 0.125 and 0.5 mg/l was 4.6 (n = 10) and 5.2% (n = 10), respectively. The interday reproducibility was 11.9% at 0.31 mg/l (n = 7) and 7.6% at 0.63 mg/l (n = 7). The lower detectable level was 0.005 mg/l.

MMC Assay in Tissular Biopsies

In seven of the patients, an immediate post-IPCH laparotomy was done to perform malignant node biopsy. The tissue fragment was weighed and kept frozen at –20 °C until assay. Biopsies were then disrupted in distilled water (6 mg biopsy/ml distilled water) at a very low temperature (–180 °C obtained by liquid nitrogen with a Spex freezer mill (Bioblock, France).

The method of Den Hartig et al. [11], which we tried to use at the beginning, did not prove to be an accurate method to measure the intratissue MMC level. So we were obliged to use other methods.

To 0.5 ml of the tissue suspension, 20 µl of vindesine 250 mg/l (Eli Lilly, France) were added as an internal standard, 0.5 ml of NaCl 0.15 *M*, $MgCl_3$ 0.05 *M* mixture. Then 2 ml of hexan were added and mixed with the suspension and then discorded to extract lipids.

Finally MMC and vindesine were extracted in 2 vol (4 ml) of chloroform: propanol 2 50:50 v:v sequentially added, mixed with the tissular suspension and separated by centrifugation.

The final volume (8 ml) was then evaporated under nitrogen stream. Then the dried residue was resolubilized in 200 µl of the mobile phase.

Chromatography of MMC and vindesine was performed using a Beckman C18 5-µm Ultrasphere 250 × 4.6 mm ID (Beckman, USA) as a column, acetonitril:methanol:(H_2SO_4 0.17 *M* with NaOH until pH 3.7) 10:20:70 v:v:v as a mobile phase flowing at 1 ml/min.

Detection of MMC and vindesine was obtained using an M 891 diode array (Waters, USA) enabling UV spectrum identification between 210 and 400 nm.

The linearity was tested from 0.4 to 52 mg/kg (mg of MMC per kg of biopsy). The extraction recovery was 65%. One reproducibility tested at 1 mg/kg showed CV of 10.3% (n = 8). The lower detectable level was 0.05 mg/kg.

Results

Figure 1 gives the kinetics of absorbtion as a percentage of the initial amount in the perfusate. A larger and rapid absorption of MMC of 32.4 ± 8.2 mg occurred during IPCH.

The MMC in serum, as described in figure 2, shows a rapid increase during IPCH, the maximum levels being attained 45 min after the beginning of IPCH. The last detectable concentration (0.006 mg/l) was obtained 6 h after the beginning of IPCH. Of the important absorbed quantity (54 ± 13.6% of the initial 60 mg), only a small amount (1.77 ± 0.33 mg) was eliminated in 24 h.

The elimination curve as established in 24 h for 10 of the patients is given in figure 3.

The MMC levels in the biopsy material are given in table 1 and compared with maximum serum levels, and the mean concentration of IPCH perfusate was 8.3 ± 7.6 mg/kg of tissue.

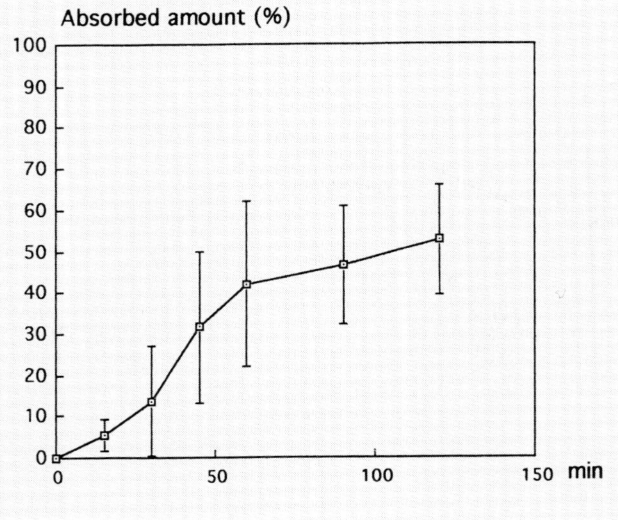

1

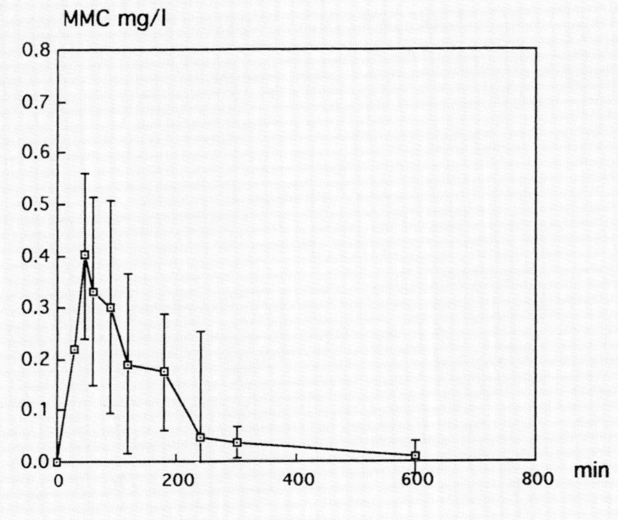

2

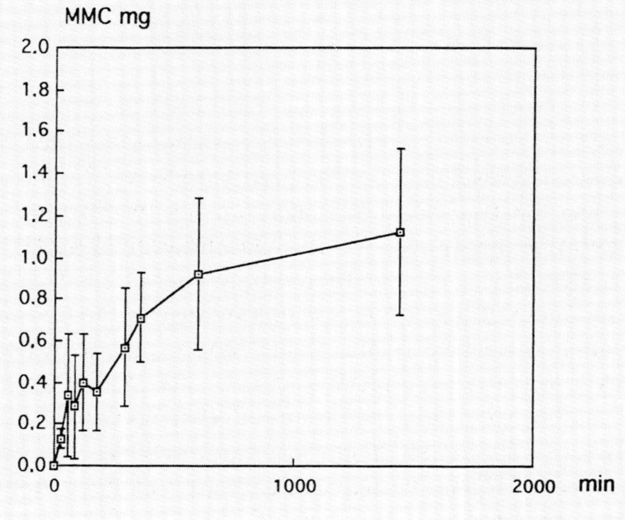

3

Discussion

Modes of administration of anticancerous drugs other than the intravenous route have been developed in recent years enabling the direct delivery of drugs to the site of action, thereby avoiding systemic toxic or adverse effects.

High-dose intraperitoneal chemotherapy with MMC is used for regional cancers to bathe tumor cells with MMC at very large extracellular concentrations unachievable by systemic administration.

Although some data are available on the pharmacokinetics of MMC during intraperitoneal chemotherapy alone [12, 13] or combined with hyperthermia [9, 10], very little information is known about the distribution, particularly in local tissues, and the long-term elimination.

During IPCH, MMC was rapidly absorbed in large amounts (32.4 ± 8.4 mg), this drug being stable at 45 °C when incubated in vitro. In spite of this important uptake, only low values were simultaneously observed in serum (maximum levels at 45 min 0.4 ± 0.18 mg/l) under those obtained after high-dose MMC intravenous infusion [14]. These data are similar to those observed previously [9], and do not vary far from those obtained by the intraperitoneal route alone [12, 13].

After IPCH, the MMC levels in serum rapidly decreased but were still detectable 6 h after beginning IPCH (0.006 mg/l). Only a small part of the absorbed amount is eliminated during 24 h (5.8 ± 3.4%) as reported after the intravenous route of administration of 6–20 mg/m^2 (8.1 ± 1.4%) [15].

The kinetics of urinary excretion are close to those previously obtained during general administration [11]: most of the MMC is excreted as soon as 10 h after the beginning.

The encouraging clinical results previously reported, the lack of general toxicity in humans [10], the good tolerance also observed in an animal model [16], the discrepancy between the large amount of MMC absorbed locally and the low levels in serum can partially be explained by an important uptake of MMC in local tissues (fig. 4).

Fig. 1. MMC absorption kinetics during IPCH.
Fig. 2. Serum levels.
Fig. 3. Urinary elimination of MMC.

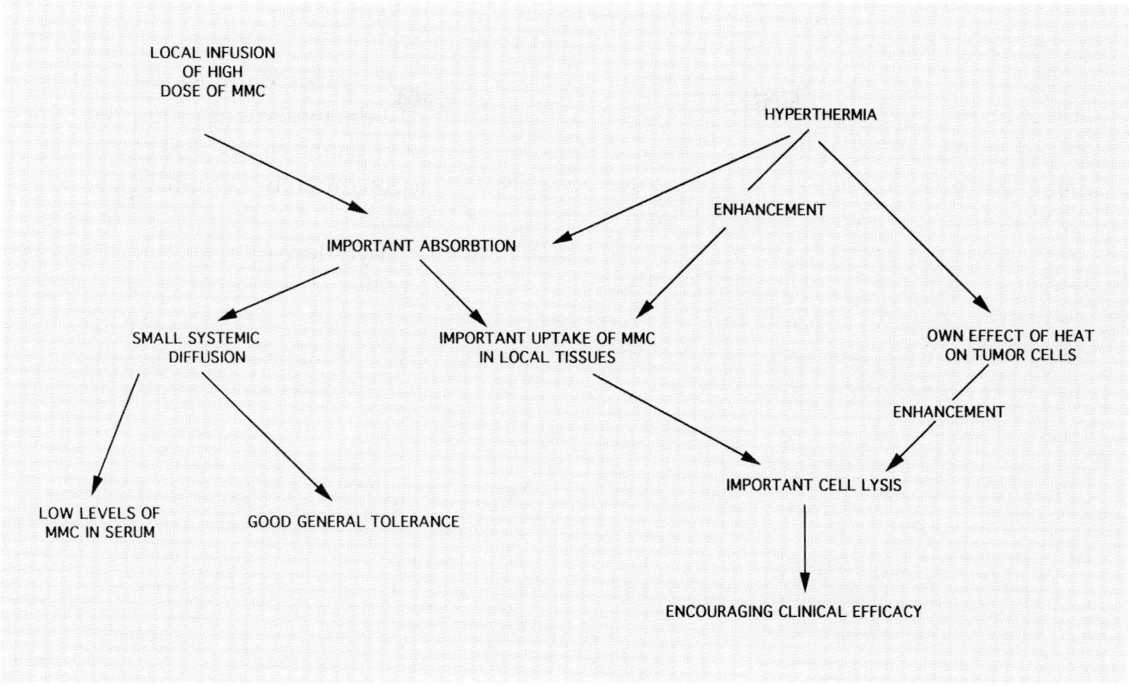

Fig. 4. Possible mechanisms of MMC action.

Table 1. MMC tissue concentration in biopsy specimens following IPCH

	Patient No.						
	1	2	3	4	5	6	7
MMC tissue levels, mg/kg	25	8.9	0.1	9.8	4.1	<2.6	7.7
Maximum MMC level in serum, mg/l	0.32	0.28	0.2	0.68	0.11	0.6	0.16
MMC levels in peritoneal fluid mg/l following IPCH	5	6.7	5	5	5	5	6.9

Biochemical data seem to confirm this hypothesis (fig. 4) only. Enzymes such as LDH and ASAT demonstrate a dramatic increase in the peritoneal fluid during IPCH, but not in the blood.

It appeared necessary to quantify the intratissue amount of MMC by means of biopsy.

First attempts to determine MMC levels in control tissue biopsies spiked with known amounts of the drug were done with the same HPLC method as for liquid samples. The necessity of a specific assay method appears rapidly: the assay device described by Den Hartig et al. [11] did not enable good and reproducible MMC extraction and chromatography due to the important binding of the drug to tissue lipids and lipoproteins and to DNA [17, 18].

This 'matrix' effect has been previously described to determine the tissue antibiotic concentration [19–21].

Many conditions were required to obtain adapted assay methods including: use of a freezer mill to obtain a good homogenate of the tissue biopsy and to protect MMC; the necessity to extract only a very small fragment of tissue; the addition of an internal standard; lipid extraction by hexan; stabilization of the DNA structure, and the MMC double-step extraction procedure, a mobile phase with a high molar concentration.

The results in milligrams of MMC per kilogram of biopsy tissue show important levels of the molecule in local tissue in spite of individual variations.

Nevertheless, MMC levels in both tissues and intraperitoneal fluids are not so different. In vivo, cells might be simultaneously bathed by an intraperitoneal MMC concentration and low blood MMC levels. Heat enhances the intracellular uptake of MMC as demonstrated in vitro by Wallner et al. [4] (fig. 4), but this drug does not strongly accumulate in cells [4]. Considering with Nix et al. [20, 21] that drugs are not homogenously distributed between the different tissue compartments (cells, interstitial fluid, blood), our in vivo data are not so different from in vitro intracellular concentration.

Hyperthermia can be expected to potentiate the in vivo action of MMC as demonstrated in vitro [4] (fig. 4). This enhancement of MMC cytotoxicity might involve an increase in the drug intracellular accumulation and activation but also alterations in DNA repair [4].

The in vivo estimation of the intracellular active metabolites as a free form and covalently or noncovalently bound to DNA [18] is still to be carried out particularly to better understand the interaction with heat, the mechanism of enhanced clinical efficiency.

References

1 Turano C, Ferraro A, Strom R: The biochemical mechanism of selective heat sensitivity of cancer cells. Eur J Cancer 1970;6:67–72.

2 Barologie B, Corry PM, Drewinko B: In vitro thermochemotherapy of human colon cancer cells with cis-dichlorodiammineplatium and mitomycin C. Cancer Res 1980;40:1165–1168.

3 Engelhardt R: Hyperthermia and drugs; in Streffer C (ed): Hyperthermia and the Therapy of Malignant Tumors. Berlin, Springer, 1987, pp 137–203.

4 Wallner KE, Banda M, Li GC: Hyperthermic enhancement of cell killing by mitomycin C in mitomycin C-resistant Chinese hamster ovary cells. Cancer Res 1987;47:1308–1312.

5 Crile G: The effect of heat and radiation on cancer implanted on the feet of mice. Cancer Res 1963;23:372–380.

6 Overgaard K, Overgarrd J: Investigations on the possibility of a thermic tumour therapy. Eur J Cancer 1972;8:65–78.

7 Hahn GM: Potential for therapy of drugs and hyperthermia. Cancer Res 1979;39:2264–2268.

8 Spratt JS, Adock A, Muskovin M, Skerill W, McKeown J: Clinical delivery system for intraperitoneal hyperthermic chemotherapy. Cancer Res 1980;40:256–260.

9 Fujimoto S, Shrestha RD, Kokubun M, Kobayashi K, Kiuchi S, Takamashi M, Konno C, Ohta M, Koikes S, Kitsukawa Y, Mitzutani M, Okui K: Clinical trial with surgery and intraperitoneal hyperthermic perfusion for peritoneal recurrence of gastrointestinal cancer. Cancer 1989;64:154–160.

10 Gilly FN, Carry PY, Sayag AC, Braillon GC, Panteix G, James IM, Cheaatsard PA, Manchon M, Desgouttes SC: Intraperitoneal cheohyperthermia (IPCH) with mitomycin C (MMC) for far advanced cancer with peritoneal seedings: Our first experience. Jpn J Hyperthermia Oncol 1991;7:208–218.

11 Den Hartig J, Van Oort WJ, Bocken MCYM, Pinedo H: High performance liquid chromatographic determination of the antitumor agent mitomycin C in human blood plasma. Anal Chim Acta 1981;127:47–53.

12 Kono A, Yashuhiro H, Tanaka M, Matsushima Y: Determination of mitomycin C in biomedical specimens by high performance liquid chromatography. J Chromatogr 1979;164:404–406.

13 Tjaden UR, De Bruun EA, Van Der Hoeven RAM, Jol C, Vander Greef J, Lingeman H: Automated analysis of mitomycin C in body fluids by high performance liquid chromatography with on-line sample pre-treatment. J Chromatogr 1987;420:53–62.

14 Schilcher BR, Young JD, Ratanatharathorn V, Karanes C, Baker LH: Clinical pharmacokinetics of high dose mitomycin C. Cancer Chemother Pharmacol 1984;13:186–190.

15 Van Hazel GA, Scott M, Rubin J, Moertel CG, Eagan RT, O'Connell MJ, Kivach JS: Pharmacokinetics of mitomycin C in patients receiving the drug alone or in combination. Cancer Treat Rep 1983;67:805–810.

16 Gilly FN, Carry PY, Sayag AC, Panteix G, Manchon M, Rochette A, Peix JL, Baulieux J, James I, Braillon G: Tolerance if intraperitoneal chemohyperthermia with mitomycin C: In vivo study dogs. Int J Hyperthermia 1992;8:659–666.

17 Orstavik J: Uptake and Release of ^{3}H-mitomycin C in mouse P388 cell cultures. Acta Pathol Microbiol Scand 1974;82(suppl):270–276.

18 Tomasy M, Mercado CM, Olson S, Chatterjee N: The mode of interaction of mitomycin C with deoxyribonucleic acid and others. Polynucleotides in vitro. Biochemistry 1974;13:4878–4887.

19 Carbon C: Significance of tissue levels for prediction of antibiotic efficacy and determination of dosage. Eur J Clin Microbiol Infect Dis 1990;9:510–516.

20 Nix DE, Goodwin D, Peloquin CA, Rotella DL, Schentag JJ: Antibiotic tissue penetration and its relevance: Models of tissue penetration and their meaning. Antimicrob Agents Chemother 1991;35:1947–1952.

21 Nix DE, Goodwin D, Peloquin CA, Rotella DL, Schentag JJ: Antibiotical tissue penetration and its relevance: Impact of tissue penetration on infection response. Antimicrob Agents Chemother 1991;35:1953–1959.

Oncology 1993;50:371–374

L. Coulon[a], *J. Bienvenu*[a]
M.C. Gutowski[a], *Y. Barbier*[b]
M. Barbier[b], *G. Panteix*[a]
F.N. Gilly[c], *P.Y. Carry*[d]
A. Sayag[d], *G. Braillon*[c]

Departments of
[a] Immunology,
[b] Radioanalysis,
[c] Surgery and
[d] Anesthesia and Intensive Care,
Centre Hospitalier Lyon-Sud,
Pierre-Bénite, France

Cytokine Production during Intraperitoneal Chemohyperthermia

Key Words
Intraperitoneal chemohyperthermia
Cytokine
Interleukin-6
Tumor necrosis factor α

Abstract
The aim of the present work is to investigate in vivo cytokine production during chemohyperthermia. 11 patients suffering from gastric adenocarcinoma (n = 6), ovarian adenocarcinoma (n = 4) or malignant mesothelioma (n = 1) were studied. Patients received 60 mg mitomycin or 100 mg cisplatin per square meter during 2 h in 6 liters of a heating solution (temperature 42 °C, flow rate 200 ml/min in a closed circuit) after previous surgical resection. Tumor necrosis factor-α (TNF-α) and interleukin-6 (IL-6) were measured at 0, 30, 45, 60 and 90 min, both in the blood stream and in the heating solution circulating intraperitoneally. We observed a slight increase in plasma IL-6 occurring as soon as 30 min, a dramatic rise in IL-6 in the heating solution. TNF-α values were only slightly augmented. In addition, the importance of various factors in the induction of IL-6 and TNF-α production during chemohyperthermia (temperature, mitomycin C, cisplatin) were studied using a whole blood ex vivo model.

Introduction

The use of intraperitoneal chemohyperthermia (IPCH) has recently become of growing interest as an anticancer therapy. In fact, IPCH seems to be a safe and reliable therapeutic manipulation for peritoneal carcinomatosis, and several groups are using this technique with good clinical results [1].

Previous studies have reported heat to be toxic in vitro for cancerous cells [2, 3]. Furthermore, the enhancement of the cytotoxicity of anticancer drugs such as mitomycin or cisplatin by heat has been shown in vitro on some malignant cells [4–6]. These properties have been verified in vivo: a favorable effect of hyperthermia on cancer implanted on the feet of mice has been described by Crile [7], and Hahn [8] observed that hyperthermia increased the cytotoxicity of anticancer drugs. Thus, these observations have led several authors to associate hyperthermia and anticancer drugs in the treatment of some tumors, particularly gastrointestinal cancers [9, 10].

Up to now, little is known about the involvement of cytokines in the effect of hyperthermia associated with chemotherapy. Nevertheless, it is likley that cytokines such as tumor necrosis factor-α (TNF-α), which is defined on the basis of its ability to cause necrosis and regression of certain tumors in vivo in mice [11] and its strong cytotoxicity on certain tumor cell lines in vitro [12], could play an important role in the efficiency of this therapy.

Dr. Jacques Bienvenu
Department of Immunology
Centre Hospitalier Lyon-Sud
F–69310 Pierre-Bénite (France)

0030–2414/93/0505–0371
$2.75/0

Table 1. Cytokine determinations in the blood of patients undergoing IPCH (n = 11, mean ± SD)

Time min	IL-6 (normal values <20 pg/ml)	TNF-α (normal values <10 pg/ml)
0	226±287	23.9±31.1
30	385±321	22.3±23.0
45	332±262	20.9±19.1
60	415±235	19.8±16.8
90	437±214	18.8±15.1

The aim of the present work is to investigate the in vivo production of cytokines (TNF-α and interleukin-6, IL-6) in patients undergoing chemohyperthermia for cancer therapy. In addition, the importance of the individual factors (i.e. temperature, mitomycin C or cisplatin) has been studied in the ex vivo production of cytokines by stimulated monocytes in a whole blood model.

Material and Methods

In vivo Study

For the in vivo study, 11 patients (1 male, 10 females) suffering from gastric adenocarcinoma (n = 6), ovarian adenocarcinoma (n = 4) or malignant mesothelioma (n = 1) were studied. IPCH was performed alone or combined with an eventual surgical excision of the primary tumor. These patients were administered 60 mg/m^2 mitomycin C or 100 mg/m^2 cisplatin, during 90 min, in 6 liters of heating solution (Dianeal glucose for peritoneal dialysis, Baxter) at 42 °C and at a flow rate of 200 ml/min.

TNF-α and IL-6 were measured both in the blood stream and the heating solution.

Samples for cytokine quantitation were drawn in EDTA tubes at 0, 30, 45, 60 and 90 min. After immediate centrifugation, plasma or heating solution supernatants were deep frozen at –80 °C.

Ex vivo Model of Cytokine Production

For this study, we used an ex vivo blood model of stimulated monocytes described by Desch et al. [13]. Blood was drawn into a heparinized tube (Kabi, Ref 82 23-12-63/9) tested for the absence of endotoxins by the chromogenic Kabitest. 450 µl of whole blood were pipetted into a 5-ml propylene sterile tube, 40 µl RPMI 1640 (control) or 40 µl mitomycin C at 0.07 mg/ml (final concentration 5.1 mg/l in the incubation medium) or 40 µl cisplatin at 0.1 mg/ml (final concentration 7.4 mg/l in the incubation medium) were added and vortexed gently. After a 20-min preincubation at 37 °C in the presence of CO_2, 50 µl RPMI 1640 (reference, no stimulation) or 50 µl lipopolysaccharide (10 ng/ml of *Escherichia coli* 055: B5 Sigma dissolved in RPMI) were introduced into the reaction medium. After mixing, the tubes were incubated for 5 h at 37, 40, 42 or 44 °C in the presence of CO_2, stopped by adding 1,500 µl of RPMI, and centrifuged for 10 min at 1,500 rpm. Supernatants were frozen in aliquots at –80 °C for cytokine determinations.

TNF-α and IL-6 quantitations were performed using the IRMA (Medgenix) and ELISA (Medgenix) techniques, respectively.

Results

In vivo Study

Blood. Although results are very variable from one patient to another, in plasma we observed a rapid but slight IL-6 increase while TNF-α values remained low (table 1). Similar results were obtained in patients undergoing elective surgery.

Heating Solution Effluent. A very important IL-6 increase was found (IL-6 4,216 ± 4,565 pg/ml at 30 min, to 9,920 ± 6,446 pg/ml at 90 min).

These data demonstrate a very important local production of IL-6 with a total amount of 50 µg IL-6 in the 6 liters of the heating solution. On the contrary, no TNF-α production was detectable in the heating solution.

These results raise a major question: is TNF-α (which is known as a thermolabile cytokine) degraded after its production by high temperature? For this reason, we carried out assays on the thermostability of TNF-α and IL-6 at 45 °C for 2 h and at 48 °C during 30 min. No degradation of IL-6 or TNF-α was observed after exposure at these temperatures.

Ex vivo Study

The importance of various factors (temperature, mitomycin or cisplatin) on cytokine production during chemohyperthermia were appraised in an ex vivo whole blood model of stimulated monocytes. Our data demonstrate that neither temperature nor mitomycin or cisplatin influence cytokine production by unstimulated monocytes. However, these factors do influence cytokine production by stimulated monocytes. Thus, TNF-α production was only slightly increased by mitomycin or cisplatin at 37 °C, but we noted an important increase at 40 °C and an even more important at 42 °C with the association of mitomycin or cisplatin. We then observed a loss of TNF-α production at 44 °C (fig. 1). Concerning IL-6, neither a temperature of less than 42 °C nor mitomycin or cisplatin influenced its production. We also noted a loss of IL-6 production at 44 °C (fig. 2). Viability assays at 44 °C by the trypan blue exclusion test verified that at this temperature no viability decrease occurred. This has already been shown by Crispen et al. [14].

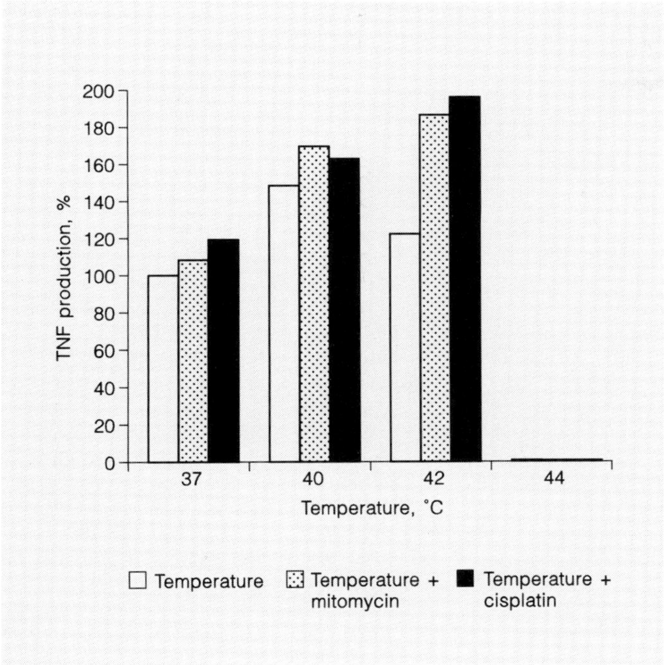

Fig. 1. Effect of temperature, mitomycin and cisplatin on TFN-α production by stimulated monocytes. Results are expressed as the percentage of TFN-α production under lipopolysaccharide stimulation at 37 °C and are the means of two experiments.

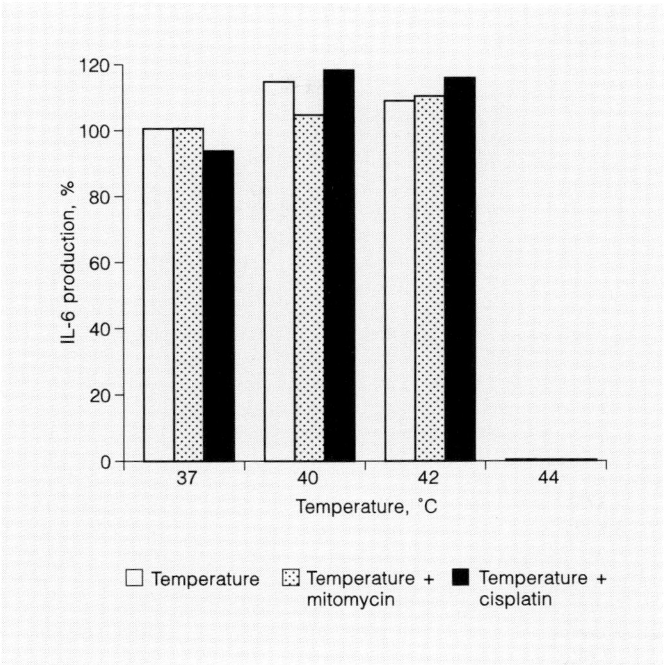

Fig. 2. Effect of temperature, mitomycin and cisplatin on IL-6 production by stimulated monocytes. Results are expressed as the percentage of IL-6 production under lipopolysaccharide stimulation at 37 °C and are the means of two experiments.

Discussion

Our work demonstrates that in vivo chemohyperthermia has little effect on cytokine production in blood, while there is an important local IL-6 production. IL-6 is the major cytokine of inflammation [15] and we believe that IL-6 increase is due to a local inflammatory reaction induced by the mechanical action, the anticancerous drugs or the temperature on peritoneal cells.

As to the ex vivo study, we noted that only TNF-α production is increased by temperature with a synergistic effect of mitomycin or cisplatin. This discrepancy between in vivo and ex vivo results could be explained in three ways: (1) cells in contact with heating solution are different from cells in the ex vivo model; (2) the ex vivo model shows an influence of temperature, mitomycin or cisplatin only on lipopolysaccharide-stimulated monocytes which is very different from in vivo conditions, and (3) it might also be that locally produced TNF-α is rapidly bound in situ by tumor cells or inhibitory factors (such as soluble receptors). If this last hypothesis is true, it could explain the good clinical results of chemohyperthermia.

The effect of in vitro hyperthermia has also been studied by Kappel et al. [16]. Using purified peripheral blood mononuclear cells, a different model from ours, they reported that elevated temperature resulted in a decreased production of IL-6 but not TNF-α. Tomasovic et al. [17] insist on the importance of appropriate temporal sequences for augmentation of either macrophage priming/triggering or monokine treatment of tumor cells, in combination with hyperthermia.

All these studies demonstrate that further investigations are needed to define the exact role of cytokines and particularly of TNF-α in chemohyperthermia. It would be of great interest to study TNF-α mRNA and TNF-α-receptor expression on biopsies of peritoneal tissue during chemohyperthermia.

References

1 Sayag AC, Gilly F, Carry PY, Panteix G, Perdrix JP, Braillon G: What about the use of intraperitoneal chemohyperthermia in abdominal and pelvic malignancies. 15th ISCH Lyon Chirurgical, Lyon, May 1992, p 41.

2 Turano C, Ferraro A, Strom R: The biochemical mechanism of selective heat sensitivity of cancer cells. Eur J Cancer 1970;6:67–72.

3 Muckle DS, Dickson JA: The selective inhibitory effect of hyperthermia on the metabolism and growth of malignant cells. Br J Cancer 1971;25:771–778.

4 Barlogie B, Corry PM, Drewinko B: In vitro thermochemotherapy of human colon cancer cells with cis-dichlorodiamine platinum and mitomycin C. Cancer Res 1980;40:1165–1168.

5 Engelhardt R: Hyperthermia and drugs; in Streffer C (ed): Hyperthermia and the Therapy of Malignant Tumors. Berlin, Springer, 1987, pp 137–203.

6 Wallner KE, Banda M, Lig C: Hyperthermic enhancement of cell killing by mitomycin C in mitomycin C resistant chinese hamster ovary cells. Cancer Res 1987;47:1308–1312.

7 Crile G: The effect of heat and radiation on cancer implanted on the feet of mice. Cancer Res 1963;23:372–380.

8 Hahn GM: Potential for therapy of drugs and hyperthermia. Cancer Res 1979;39:2264–2268.

9 Fujimoto S, Shrestha RD, Kokbun M, Kobayashi K, Kiwhi S, Takahasui M, Konno C, Ohta M, Koike S, Kitsukaway Y, Mitzutani M, Okui K: Clinical trial with surgery and intraperitoneal hyperthermic perfusion for peritoneal recurrence of gastro intestinal cancer. Cancer 1989; 64:154–160.

10 Gilly FN, Carry PY, Sayag A, Braillon G, Panteix G, James I, Cheateasard P, Manchon M, Desgoutte S: Intraperitoneal chemohyperthermia (IPCH) with mitomycin C (MMC) for far advanced cancer with peritoneal seedings: Our first experience. Jpn J Hyperthermic Oncol 1991;7:208–218.

11 Carswell E, Old L, Kassel R, Green S, Flore N, Willarmson B: An endotoxin induced serum factor that causes necrosis of tumors. Proc Natl Acad Sci USA 1975;72:3666–3670.

12 Sugerman BJ, Aggarwal BB, Hass PE, Figari IS, Palladino M, Shepard HM: Recombinant human tumor necrosis factor-alpha: Effects on the proliferation of normal and transformed cells. Science 1985;230:943–945.

13 Desch CE, Kovach NL, Present W, Broyles C, Harlan JM: Production of human tumor necrosis factor from whole blood ex vivo. Lymphokine Res 1989;8:141–146.

14 Crispen RG, Grand L, Spahic O, Kim YD, Levin RD, Williams RM: Effect of whole body hyperthermia on the immune system (abstract). 13th Meet Int Clin Hyperthermia Soc, Acre, May 1990.

15 Hirano T, Akira S, Taga T, Kishimoto T: Biological and clinical aspects of interleukin-6. Immunol Today 1990;11:443–449.

16 Kappel M, Diamant M, Hansen MB, Klokker M, Pedersen BK: Effects of in vitro hyperthermic on the proliferative response of blood mononuclear cell subsets, and detection of interleukin 1 and 6, tumour necrosis factor-alpha and interferon-gamma. Immunology 1991;73: 304–308.

17 Tomasovic SP, Barta M, Klostergaard J: Temporal dependence of hyperthermic augmentation of macrophage-TNF production and tumor cell-TNF sensitization. Int J Hyperthermia 1989;5:625–639.

Oncology 1993;50:375–379

Alain Sibille
Frédéric Prat
Jean-Yves Chapelon
Fatima Abou El Fadil
Luc Henry
Yves Theillère
Thierry Ponchon
Dominique Cathignol

Department of Digestive Disease and INSERM U 45, Edouard-Herriot Hospital, and INSERM U 281, Lyon, France

Extracorporeal Ablation of Liver Tissue by High-Intensity Focused Ultrasound

Key Words
High-intensity focused ultrasound
Extracorporeal
Thermal effect
Liver tissue
Liver tumors

Abstract
High-intensity focused ultrasound (HIFU) may produce a well-delineated lesion of coagulation necrosis in deep organs, by means of an extracorporeal transducer. Applications of this method to the liver in animal models have been studied for many years. The effects of HIFU on the normal liver parenchyma and on hepatic tumors are reviewed. In the normal rabbit liver in vivo we showed the relation between intensity levels and exposure times and the need to adapt intensity to the depth of the target. No severe complications were observed when an intensity of 1,000 W/cm^2 was used. HIFU is a noninvasive method for the local destruction of liver tumors. In experimental models, safety and efficacy were demonstrated. HIFU may be interesting for the treatment of some human liver tumors.

Conventional hyperthermia (40–45 °C) does not cause tissue damage by itself, but tumor tissue is known to be more susceptible to heat than normal tissue. This effect has been used to potentiate other therapeutic means such as radiotherapy or chemotherapy. Hyperthermia may be produced by low-intensity ultrasound, microwaves, high-frequency current or perfusion of heated solutions. Irreversible enzymatic changes, collagen denaturation and coagulation necrosis of tissue are achieved for temperatures as high as 60–100 °C. Such temperatures are used for coagulation, resection or destruction of tissues. At temperatures above 100 °C, vaporization and carbonization of tissue occur. High-intensity focused ultrasound (HIFU) is one means of producing coagulation necrosis that allows extracorporeal treatment.

Principle of HIFU

HIFU may produce two kinds of effect. First, when the ultrasound wave is absorbed at the focal area, it is partially converted into heat causing a temperature rise around 80 °C. This thermal effect provokes tissue destruction by coagulation necrosis. Because the temperature rise occurs quickly and briefly, there is no heat diffusion out of the focus area and no damage is caused in the surrounding tissues.

Second, a mechanical effect, cavitation, can occur. In biological tissues, with the depression induced by the negative part of the ultrasound wave, intracellular water may enter the gaseous phase. That leads to the development of microbubbles. These bubbles suddenly collapse when

We thank Mr. E. Blanc from Technomed International for his technical support.

F. Prat
INSERM U 281
151, Cours Albert-Thomas
F-69003 Lyon (France)

0030–2414/93/0505–0375
$2.75/0

they reach their size of resonance, producing shock waves of very high pressure (20–30,000 bars) [1] damaging surrounding cells.

Thermal damage is usually assumed to occur with long exposure times (above 1 s) and relatively low intensities (below 1,000 W/cm^2) [2], while cavitation occurs with high peak intensities and brief exposure periods [3]. There is a range of levels of intensities and exposure periods for which both phenomena may not be distinctly identified and probably occur simultaneously.

HIFU are generated by focusing piezoelectric transducers (lens-focused quarz transducer [4], piezocomposite shell [5], multiple piezoelectric ceramics [6]) connected to a power amplifier and a pulse generator.

Transducer and target must be placed in a tank filled with degassed and deionized water in order to allow good ultrasonic transmission. An imaging system is needed to localize the target. A computer is required to plan and monitor irradiation of the target.

Experimental Studies

The use of HIFU to induce tissular necrosis has been studied for many years on different organs (liver, kidney, brain, eye, testes) in animals and in humans [7–15].

These studies have shown that it was possible to induce well-delineated lesions corresponding to the focal area of the ultrasound beams generated by the HIFU transducer, in deep organs without causing damage to the more superficial tissues.

Chapelon et al. [5] showed that the lesion in the rat kidney may be changed from coagulation necrosis to 'punched out' cavity by increasing the intensity and/or shortening the exposure time. The effects of HIFU on tumors were also studied. The same team from our laboratory obtained complete destruction and no local recurrence in 16 of 49 rats (32%) subcutaneously injected with the Dunning-Mat-Ly-Lu, a highly treatment-resistant tumor. 14% were alive and metastases-free 12 months after the treatment. With the less-aggressive AT2 subline, they obtained 96% complete tumor necrosis and a disease-free survival rate of 64% at 115 days [14].

HIFU has also been applied to humans. Fry [15] treated patients with hyperkinetic disorders by application of HIFU to the brain tissue. More recently, Vallancien et al. [6], in a feasibility study, treated patients with benign prostatic hyperplasia, superficial bladder tumor or renal pathology with HIFU prior to surgery. Although they did not prove the efficacy of the method from the clinical point of view, they obtained few side effects, mostly skin burns.

For the good penetration of ultrasound in tissue and the sharp limits of the lesions so obtained, HIFU applications to the liver are potentially interesting.

Applications to the Liver

Liver Tissue. There are several reports in the literature concerning the use of HIFU in liver tissue in different experimental models [7, 9, 16], in vivo or in vitro. The ability of HIFU to cause delineated irreversible tissue destruction in the liver has been well documented.

Taylor and Connolly [7] in 1969 showed that the effects of HIFU depended on the temporal mode of delivery of the ultrasound (pulse duration and interval of time between two pulses). He also described lesions due to the reflection of ultrasound waves at the liver-air interfaces. Linke et al. [9] observed a maximum lesion size the third day after the treatment, and its replacement by a thin fibrous scar after several months. In a quantitative and qualitative study, using a 1.8-MHz transducer on porcine liver in vitro, Ter Haar et al. [16] showed that with increasing intensity at constant exposure time the lesion volume increased but was developing toward the liver surface, in front of the focal plane. When intensity was kept constant and exposure time increased, the lesion size increased to a maximum after which longer exposure times did not cause larger lesions.

We conducted several experiments with extracorporeal HIFU (1 MHz transducer) on rabbit liver in vivo. In preliminary studies, we confirmed the possibility to produce a well-delineated lesion deep in the liver. It was a lesion of coagulation necrosis, surrounded by an inflammatory reaction. Then we wanted to assess the relationship between pulse intensity and duration on the one hand and the depth of the target in the liver on the other hand for a given volume of tissue destruction. In a first group of rabbits we tested different parameters in order to define the best level of intensity and exposure time to obtain an optimal lesion. An optimal result was reached when the necrosis area in the liver had the same dimensions as the desired volume at the desired location. Increasing intensities were used with different pulse durations. An optimal result was obtained for each intensity level at a specific exposure time: the higher the intensity the shorter the time. In a second group of rabbits, different depths were targeted in the liver. We observed that for a constant result, intensities had to be increased according to the law of sound absorption in tissues.

We also analyzed the features of the lesion 4 weeks after treatment. We observed a fibrous scar surrounding the necrosis when a large volume was treated, and only a thin fibrous scar when a smaller volume was treated.

Hepatic Tumors. Effects of HIFU on hepatic tumors were also studied. With a 4-MHz transducer Moore et al. [17] treated Morris hepatomas injected subcutaneously in rats. They used intensities of 400 W/cm^2. Tumor volume in HIFU-treated rats was significantly smaller than in controls at 4 weeks. A synergism bertween HIFU and chemotherapy (intraperitoneal cyclophosphamide) was also observed.

Yang et al. [4], from the same team, treated Morris hepatomas implanted in the liver of rats. For HIFU treatment (4 MHz, 550 W/cm^2), the tumor-bearing livers were exteriorized after laparotomy. Tumor growth inhibition rates from 65 to 93% were seen in the HIFU-treated group. They also showed a longer median survival in groups treated with HIFU or HIFU combined with doxorubicin.

Ter Haar et al. [18] used HIFU to treat a cat with a metastatic melanoma infiltrating the liver, in vivo. A defined region of the liver tumor was successfully treated and no viable cells were found in that volume.

Our group was the first to describe extracorporeal HIFU treatment in vivo in a group of rabbits with liver tumor in situ [19]. Destruction rates as estimates by quantitative microscopy on millimetric tissue slices ranged from 42.5 to 100%. Histology showed homogenous and well-delineated coagulation necrosis corresponding to the target volume in the depth of the liver. No viable tumor tissue remained in the treated area.

Side Effects of HIFU

Several side effects associated with HIFU were reported in the literature. Fry and Johnson [11] in 1978 observed severe skin burns. Moore et al. [17], with a 4-MHz transducer, reported full thickness skin burns at energy levels of 500–700 W/cm^2. Even for lower energy depositions (400 W/cm^2), 4 animals were burnt. Because he worked with a subcutaneously implanted tumor, the fact that the skin was close to the focus area may explain this high rate of skin burns at relatively low intensities.

In our series, we observed skin burns in preliminary studies. Convenient adaptation of intensities and pulse durations as well as taking into account the law of ultrasound absorption made it possible to get rid of such side effects.

Yang et al. [4] observed a slight elevation of the ASAT during the first 24 h after the treatment. Bowel perforation was induced by HIFU in 2 rats who died of this complication. The authors explained these dismal events by the small size of the rat liver.

Indeed, in our study, such complications occured only with high levels of energy and large target volumes. Two groups of rabbits treated with different parameters were studied to analyze the side effects of HIFU. Autopsy was performed on day 28. Rabbits treated with an intensity of 1,000 W/cm^2 and a target volume of $10 \times 10 \times 10$ mm did not present any significant alteration of the blood parameters or severe side-effects. In one, the stomach wall was affected by the HIFU: a scar adherent to the liver was seen but there was no perforation. In rabbits treated with an intensity of 3,000 W/cm^2 and a larger target volume ($14 \times 14 \times 20$ mm) blood samples showed an increase of the transaminase rate on day 2 which was normalized on day 28. Autopsy revealed a perforation of the diaphragm in 50% of the cases and a gastric perforation in 25%. Such complications seem to be related to the size of the target volume: with a large target area, some of the volume may be more frequently located outside of the liver so damaging adjacent organs.

A controversial question is the possibility of increased risk of metastases after the tumor has been disrupted by HIFU. In a report by Fry and Johnson [11], a higher rate (17–44%) of metastases was observed after HIFU treatment than in the control group. However, that feature was not found in other studies. On the other hand, Yang et al. [4] described a lower rate of lung metastases in the HIFU-treated group. In the model of prostatic adenocarcinoma (Dunning-AT2), Chapelon et al. [14] found metastases in 16% of the treated rats as compared to 28% in the control group. Other studies will be needed to address that question.

Imaging of the HIFU-Treated Liver

Interestingly, Ter Haar et al. [16] observed that the HIFU-induced lesion appeared as a hyperechogenic pattern on a real-time ultrasound scanner. However, since it was an in vitro study on excised pig liver, ultrasound imaging may have been different than in the case of an in vivo model. In our in vivo study, ultrasound imaging was performed immediately after treatment. After treatment of normal rabbit livers we seldom observed hyperechogenic pattern at the target location, even when a homogenous lesion was observed at autopsy. When this hyperechogenic pattern was present immediately after the treatment, it disappeared 3 weeks later. In contrast, the hyperechogenic pattern was always present after successful treatment of rabbits bearing a liver tumor. However, there was no correlation between the size of the echogenic area and the volume of the coagulation necrosis at autopsy.

Magnetic resonance imaging (MRI) may be more specific. MRI is expected to distinguish between tumor, spontaneous tumor necrosis and HIFU-induced coagulation necrosis. At present, there is no study on the MRI of the HIFU lesion in the liver.

Special Topics

Some interesting features of HIFU have been reported in addition to the direct destructive effect on tumors.

Yang et al. [20] treated mice inoculated with C1300 neuroblastoma in the flank; after HIFU treatment or a sham procedure, they reinoculated these mice with neuroblastoma cells in the opposite flank. Tumor growth following that second challenge was significantly reduced in the group of HIFU-cured animals. Despite the fact that immunostimulation by tumor breakdown products may be a tumor-specific feature, a HIFU effect is putative. Further studies must be done to test this hypothesis.

In another experiment, researchers [20] investigated the effects of HIFU on the large abdominal vessels in rabbits. While the lumbar muscle was destroyed, the vessels remained poorly affected by the HIFU treatment. This is in agreement with our results. In three rabbits, HIFU were intentionally applied to a large intrahepatic vessel (portal branch). They were treated with high intensity and large target volume. Doppler examination and autopsy at 4 weeks did not show any vascular thrombosis or disruption. The lesser susceptibility of vessels to HIFU may be due to heat evacuation by the blood flow, thus limiting the temperature rise.

Is it Possible to Treat Human Liver Tumors with HIFU?

The potential interest of the applications of HIFU to the liver is to treat hepatic tumors. Indeed, hepatocellular carcinoma and liver metastases of various types of cancers are frequent causes of death [21, 22] and curative treatments are lacking. Surgical resection may be curative but is only possible in a minority of cases [23]. Nonsurgical methods such as chemoembolization, percutaneous ethanol injection or heating the tumor with laser probes are under investigation [24]. HIFU could be a new particularly poorly-invasive option. It could be used alone or in combination with other techniques. However, at the present time, some difficulties must be considered. (1) Therapeutic parameters may greatly differ from animal to man. (2) Indications must be limited to tumors accessible with an ultrasound scanner. (3) Bones and lungs oppose the penetration of ultrasound and some areas of the liver parenchyma may be difficult to reach with the focused beam. (4) The size of tumors must be taken into account: in large tumors, several sessions will be necessary and we do not know how large the target volume may be without increasing the risk of complications.

Conclusion

Extracorporeal HIFU may induce intrahepatic, well-delineated lesions in the depth of the liver in animal models in vivo. Quantitative and qualitative studies helped to define the modalities of treatment. First experiments of intrahepatic tumor destruction in animals are encouraging. However, further studies are needed before application of HIFU to humans with liver cancer, including analysis of the metastasis rate and long-term survival in animals treated by HIFU. Some advantageous features such as immunostimulation and the poor susceptibility of vessels must be more documented, being potentially interesting for the treatment of liver cancer.

References

1 Flynn HG: Generation of transient cavities in liquids by microsecond pulses of ultrasound. J Acoust Soc Am 1982;72:1926–1932.
2 Frizzell LA, Linke CA, Cartensen EL, Fridd CW: Threshold for local ultrasonic lesions in rabbit kidney, liver and testicle. IEEE Trans Biomed Eng 1977;24:393–396.
3 Dunn F, Fry FJ: Ultrasonic threshold dosages for the mammalian central nervous system. IEEE Trans Biomed Eng 1971;18:253–256.
4 Yang R, Reilly CR, Rescorla FJ, Faught PR, Sanghvi NT, Fry FJ, Franklin TD, Lumeng L, Grosfeld JL: High-intensity focused ultrasound in the treatment of experimental liver cancer. Arch Surg 1991;126:1002–1010.
5 Chapelon JY, Margonari J, Theillère Y, Gorry F, Blanc E, Gelet A: Effects of high energy focused ultrasound on kidney tissue in the rat and the dog. Eur Urol 1992;22:147–152.
6 Vallancien G, Harouni M, Veillon B, Mombet A, Prapotnich D, Brisset JM, Bougaran J: Focused extracorporeal pyrotherapy: Feasibility study in man. J Endourol 1992;6:173–181.
7 Taylor KJW, Connolly CC: Differing hepatic lesions caused by the same dose of ultrasound. J Pathol 1969;98:291–293.
8 Fry FJ, Kossoff G, Eggleton RC, Dunn F: Threshold ultrasonic dosages for structural changes in the mammalian brain. J Acoust Soc Am 1970;48:1413–1417.
9 Linke CA, Carstensen EL, Frizzell LA, Elbadawi A, Fridd CW: Localized tissue destruction by high intensity focused ultrasound. Arch Surg 1973;107:887–891.
10 Lizzi FL, Coleman DJ, Driller J, Franzen LA, Jakobiec FA: Experimental, ultrasonically induced lesions in the retina, choroid and sclera. Invest Ophtalmol Visual Sci 1978;17:350–360.
11 Fry FJ, Johnson LK: Tumor irradiation with intense ultrasound. Ultrasound Med Biol 1978; 4:337–341.
12 Vallancien G, Chopin D, Thibauld P, D'Avila C, Guillonneau B, Perreira E, Veillon B, Brisset JM, Bougaran J: Pyrothérapie extra-corporelle focalisée: Premiers résultats expérimentaux. Prog Urol 1991;1:151–155.
13 Chapelon JY, Margonari J, Cathignol D, Gelet A: Evaluation of tumor ablation with focused ultrasound in the treatment of Dunning tumors. Proc IEEE Eng Med Biol Soc 1991;13: 177–178.
14 Chapelon JY, Margonari J, Vernier F, Gorry F, Ecochard R, Gelet A: In vivo effects of high-intensity ultrasound on prostatic adenocarcinoma Dunning R3327. Cancer Res 1992; 52; 6353–6357.
15 Fry FJ: Biological effects of ultrasound: A review. Proc IEEE 1979;67:604–619.
16 Ter Haar G, Sinnett D, Rivens I: High intensity focused ultrasound: A surgical technique for the treatment of discrete liver tumours. Phys Med Biol 1989;34:1743–1750.
17 Moore WE, Lopez RM, Matthews DE, Sheets PW, Etchison MR, Hurwitz AS, Chalian AA, Fry FJ, Vane DW, Grosfeld JL: Evaluation of high-intensity therapeutic ultrasound irradiation in the treatment of experimental hepatoma. J Pediatr Surg 1989;24:30–33.
18 Ter Haar GR, Clarke RL, Vaughan MG, Hill CR: Trackless surgery using focused ultrasound: Technique and case report. Min Inv Ther 1991;1:13–19.
19 Chapelon JY, Prat F, Sibille A, Abou El Fadil F, Henry L, Theillère Y, Cathignol D: Extracorporeal selective focused destruction of hepatic tumours by high intensity ultrasound in rabbits bearing VX-2 carcinoma. Min Inv Ther 1992; 1:287–293.
20 Yang R, Reilly CR, Rescorla FJ, Sanghvi NT, Fry FJ, Franklin TD, Grosfeld JL: Effects of high-intensity focused ultrasound in the treatment of experimental neuroblastoma. J Pediatr Surg 1992;27:246–251.
21 Di Bisceglie AM, Rustgi VK, Hoofnagle JH, Dusheiko GM, Lotze MT: Hepatocellular carcinoma. Ann Intern Med 1988;108:390–401.
22 Gastrointestinal Tumour Study Group: Adjuvant therapy of colon cancer: Results of a prospectively randomized trial. N Engl J Med 1984;310:737–743.
23 Bruix J, Cirera I, Calvet X, Fuster J, Bru C, Ayuso C, Vilana R, Biox L, Visa J, Rodes J: Surgical resection and survival in Western patients with hepatocellular carcinoma. J Hepatol 1992;15:350–355.
24 Hobbs KEF, Dusheiko GM: Management of hepatocellular carcinoma. J Hepatol 1992;15: 281–283.

Oncology 1993;50:380–389

Milton B. Yatvin[a]
Michael H.B. Stowell[b]
Corklin R. Steinhart[c]

[a] Division of Radiation and Thermal Biology, Department of Radiation Oncology, Oregon Health Sciences University, Portland, Oreg.;
[b] Department of Chemical Biology, California Institute of Technology, Pasadena, Calif.;
[c] Special Immunology Unit, Department of Medicine, Mercy Hospital, Miami, Fla., USA

Shedding Light on the Use of Heat to Treat HIV Infections

Key Words
Hyperthermia
HIV
Immune enhancement
Stress proteins

Abstract
Considering the lack of effectiveness of current therapies to treat HIV disease, the authors present observations that provide a strong cogent argument for critically designed and meticulously performed clinical trials employing whole body hyperthermia with or without other therapeutic modalities. Only as a result of such clinical trials will it be possible to fairly evaluate the role of hyperthermia as a potential therapy for treatment of chronic HIV infection.

Introduction

Although there has been considerable progress made in the management of chronic HIV infection, at present no effective treatment exists. The current cornerstone of therapy is zidovudine (3'-azido-2',3'-dideoxythymidine), for which a number of positive effects have been reported. At best, however, the positive responses are ephemeral. The same appears to hold true for the two newer retroviral agents 2',3'-dideoxyinosine and 2',3'-dideoxycytidine. An approach to therapy which could possibly have longer lasting effects and which has not been fairly evaluated in the past is whole body hyperthermia (WBH).

Since WBH was first proposed as part of a combined modality approach for the treatment of AIDS [1, 2], negative responses, both prior and subsequent to its clinical use, have mitigated against its receiving a fair and impartial evaluation. Seemingly damning arguments against its use emanated from a special NIH panel created by the National Institute of Allergy and Infectious Diseases to 'evaluate' the use of WBH in AIDS therapy. In August of 1990, the panel issued its report based on their evaluation of the efficacy of single WBH treatments that were performed on 2 patients. Reportedly a WBH temperature of 42 °C was obtained by heating blood to 45 °C extracorporeally and recirculating it into the patient's body. Using this information the panel reached the very broad conclusion that '– there appears to be no clinical, immunological or virological support for the use of hyperthermia in the treatment of HIV disease or Kaposi's sarcoma. Neither does there appear to be any support for further human experimentation in this area at this time'.

As part of its 'rationale' for coming to the above conclusions the panel cited a study by Stanley et al. [3], from the laboratory of Anthony Fauci, Director, National Institute of Allergies and Infectious Diseases (NIAID). The study, which apparently had great impact on the panel, indicated that heat shock upregulates the production of HIV in chronically infected human cells cultured in vitro. Along the same lines, they noted 'that a common sequela of systemic hyperthermia is activation of latent herpes simplex virus type 1'. From the above study and the clini-

Milton B. Yatvin
Division of Radiation and Thermal Biology
Department of Radiation Oncology
Oregon Health Sciences University
Portland, OR 97201-3098 (USA)

0030–2414/93/0505–0380
$2.75/0

cal observation they inferred that treating HIV-infected people with hyperthermia would be detrimental because it would increase virus production.

Statements by Dr. Fauci, who was not a member of the panel, appeared in a New York Times article [4], further calling into question the use of WBH to treat HIV infection. In the article Dr. Fauci was quoted as follows: 'There is very little rationale for the procedure, since some laboratory experiments have shown that heat actually increased the activity of the virus rather than killing it; and there is substantial risk from the procedure, including blood clotting and heart failure, which can be caused by the heating and cooling procedures.' Unfortunately, the panel's report and the subsequent public statements, although predicated to a great extent either on misinformation or disputable interpretations of salient facts, were widely disseminated, thereby virtually closing the door on a carefully controlled fair evaluation of WBH as a treatment modality for HIV infection.

In the following discussion we intend to address the issues raised initially by the NIH special panel, and subsequently by NIH personnel in the press, and to provide additional experimental, clinical and theoretical support for the use of WBH to treat HIV disease.

Experimental and Theoretical Evidence Which Supports the Use of WBH to Treat HIV Infections

Role of Cellular Components and Heat in Death of Virally Infected Cells

The panel attempted to refute Yatvin's [2] arguments favoring the use of WBH to treat HIV infections by asserting that 'Yatvin's theoretical arguments do not address the fact that following acute HIV infection, the virus is incorporated into human DNA and thus is not accessible to heat denaturation ... while HIV is a relatively heat-sensitive virus, it does not become denatured by heating to 42 °C, the maximum core temperature tolerated during human systemic hyperthermia.' It is surprising that the panel and its hyperthermia expert, Dr. H.I. Robins, failed to consider available information indicating that both acute and chronic viral infection might compromise a cell's ability to withstand a heat insult. In fact, Robins's apparent rationale for treating cancer patients with WBH, either alone or in combination with drugs and/or biological response modifiers, has been the potential therapeutic benefit that might be gained by preferentially destroying malignant cells while sparing normal cells [5, 6].

Wong et al. [7] and Omar et al. [8] report that virally infected cells are more heat sensitive than uninfected cells. Mouse embryo cells infected with SV40 were 20-fold more sensitive to heat killing (44 °C for 3 h) than uninfected cells [7]. Following exposure for 5 h to 42 °C, twice as many HUT-78 cells (human T-cell line) infected with HIV were killed as noninfected cells; after 10 h of heating, all the infected cells were killed while 50% of the noninfected cells survived [9]. The report of Stanley et al. [3] failed to note that Wong et al. [7] had reported differential heat killing of HIV-infected cells. Nor did they draw attention to their own results showing that 40–60% of HIV-infected ACH-2 T-lymphocytic cells were dead by day 2 after being exposed to 42.6 °C for 1 h!

There are several possible reasons for these results. Infected cells reportedly contain large quantities of unintegrated viral DNA, heterodisperse RNAs and viral core proteins. Shaw et al. [10] found that a substantial amount of HIV DNA persists in cells freshly infected with HIV, chronically infected cells and biopsy tissue from lymph nodes of AIDS patients. It was present not only in the unintegrated linear double-stranded form but also in the closed and nicked circular form, and its presence has been correlated with the cytopathic effects of some animal retroviruses [11, 12]. Koga et al. [13] found that high levels of aberrant heterodisperse RNAs accumulate in both HIV-infected cells and in normal lymphocytes. Somasundaran and Robinson [14] also reported high copy numbers of viral RNA in HIV-expressing cells and found that as much as 40% of the infected cell's protein synthetic activity was devoted to synthesis of viral *gag* proteins. These high levels of unintegrated DNA and viral RNA and protein interfere with cell function and are likely to predispose such cells to heat inactivation.

At the time the panel issued their report, relevant information existed in the literature regarding the effect of hyperthermia on the Golgi apparatus in virally infected and transformed cells which they overlooked. Clark et al. [15], whose group included Dr. Robins, observed vesiculation of the Golgi immediately after local in vivo exposure of mammary tumor cells to 43.5 °C for 1 h, which persisted for as long as 30 h.

Recently we have used fluorescent-labeled lipids to study lipid distribution and processing in heated virus-infected cells. The strikingly altered, vesiculated Golgi morphology of virus producing D17 cells exposed to fluorescent ceramide is depicted in figure 1.

These dog osteosarcoma cells, chronically infected with avian spleen necrosis retrovirus, were incubated in the presence of 2.5 μM NBD-ceramide (NBD-CER) for

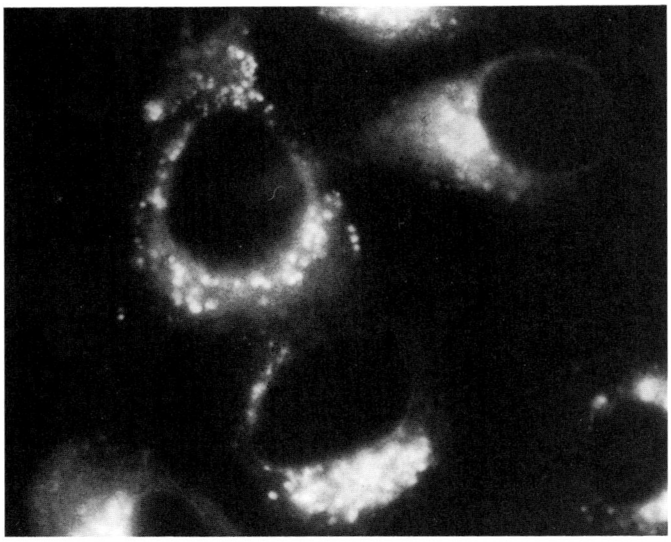

Fig. 1. Dog osteogenic sarcoma cells chronically infected with spleen necrosis virus were grown in 2.5 μ*M* NBD-CER, fluorescent ceramide, at 37 °C for 24 h, rinsed with medium and photographed under UV light in a fluorescent microscope using a blue filter (490 nm). The presence of NBD-CER did not affect cell growth. × 100.

24 h. Our results confirmed earlier observations by others that both Golgi appearance and function are markedly altered in virus-infected cells and tumor cells [16, 17].

In an attempt to determine whether infected cells were predisposed to heat injury, we studied the intracellular distribution and processing of fluorescently labeled ceramide in heated control and chronically virus-infected cells. The cells used were a murine NIH/3T3 cell line derivative which produces a replication-defective ecotropic retrovirus [18]. The uninfected cells were NIH/3T3 cells. NBD-CER, which concentrates in and clearly defines the Golgi apparatus in metabolizing cells, moved more slowly out of the Golgi in virally infected cells. Such was the case for both heated and unheated cells. Within 8 h after heating, Golgi labeling in the heated 3T3 cells was indistinguishable from that of unheated controls, but virus-infected cells required at least 24 h before returning to their preheat appearance.

A possible reason for the increased heat sensitivity of infected cells is that virus-budding damages cellular membranes, making them more susceptible to heat damage. The plasma membranes from both acutely [19–21] and chronically virus-infected cells have increased permeability [21, 22], which allows various proteins that inhibit HIV replication to enter [19, 21, 22]. Because most proviral DNA containing CD4+ T cells are latently infected and do not produce either HIV of HIV proteins, their plasma membranes will not show increased permeability. However, when chronically infected latent cells are stimulated to express virus and viral proteins, cell membrane damage occurs as a result of viral budding [23]. Other researchers have shown that the influx of divalent cations into infected human T cells is greatly accelerated, but synthesis of phosphatidylcholine and diacylglycerol is reduced [24]. The latter lipid serves as a second messenger in addition to being an intermediate in phospholipid synthesis. These observations indicate that it is quite likely that both acutely and upregulated chronically virus-infected cells will be more heat sensitive than uninfected cells. This view is supported by numerous studies indicating that heat-induce changes in membranes, particularly membrane lipids, are the primary event in heat-induced cell damage and inactivation [for reviews, see Yatvin et al., 25, and Yatvin and Cramp, 26].

Heat and Viral Stability

The panel's statement '... while HIV is a relatively heat-sensitive virus it does not become denatured by heating at 42 °C ...' ignored both the realities of how WBH was induced and what Yatvin [2] stated in his publication. Although HIV is relatively stable at 37 °C, it is relatively rapidly inactivated at 42–45 °C [27], and heat inactivation of retroviruses is not only contingent on the temperature attained, but the duration of heating [28]. When blood is heated extracorporeally, temperatures at least as high as 45 °C can be attained. Depending on the rate of blood flow, even higher extracorporeal temperatures can be safely attained with the blood of patients.

Furthermore, as suggested by Yatvin [2], heat inactivation will be enhanced when it is administered in conjunction with agents such as butylated hydroxytoluene, adamantane, ethanol, or local anesthetics such as lidocaine or other chaotropic agents. This class of chaotropic agents has been shown to rapidly inactivate lipid-containing viruses either alone or in combination with heat [29–31]. Such an effect is shown in figure 2, in which Moloney murine leukemia virus was heated in the presence of different ethanol concentrations.

Another way of modifying viral infectivity is the differential removal of cholesterol from viral envelopes, a method which has been shown to reduce the infectivity of vesicular stomatitis virus. Cholesterol-rich domains are an important component of the HIV envelope and probably play an important role in maintaining viral infectivity [32]. The use of inhibitors of cholesterol metabolism in HIV-infected persons may be a way of accomplishing

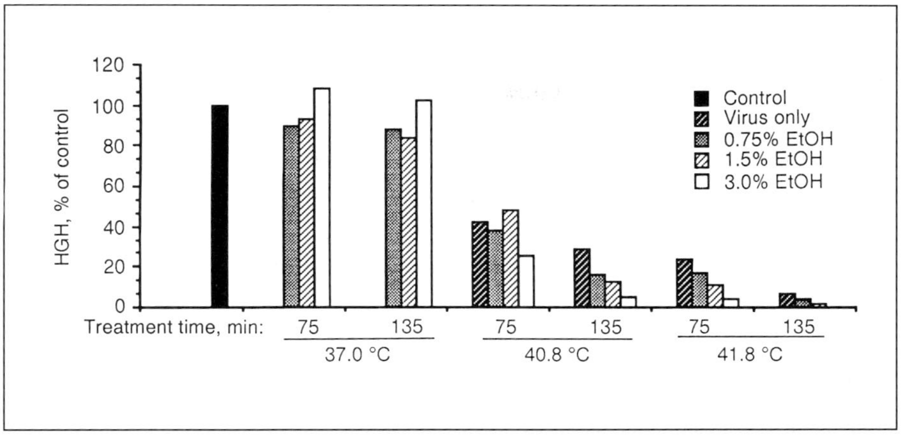

Fig. 2. Human growth hormone (HGH) content of cells exposed to virus in the presence of EtOH. Cell-free virus was heated in DMEM plus 10% fetal bovine serum containing 0–3% ethanol (EtOH) for either 75 or 135 min at various temperatures up to 41.8 °C. The virus used contains the human growth hormone gene as a reporter. Unheated cells were exposed to the virus preparation for 2 h, rinsed, incubated for 36 h and assayed for HGH by radioimmunoassay. The effects of temperature, heating interval and EtOH concentration of the virus are clearly demonstrated.

viral envelope modification. HIV has a high cholesterol content relative to its phospholipid content [33], and it is possible that depleting cholesterol from HIV could increase its sensitivity to heat [27].

Heat Prevention of Accidental HIV Infection

Shown in figure 3 are infection rates obtained in cells exposed to heat at various times after infection. When cells are heated to 42.8 °C immediately after or up to 48 h following a 2-hour exposure to virus, the rate of infection varies with the time of heating. Up to 6 h after exposure, heat markedly reduces the rate of infection.

An important aspect to be considered when employing hyperthermia in cases of accidental exposure to infected body fluids is its ability to influence subsequent infection in cells previously exposed to heat. Prior heating of cells (42.8 °C for 135 min) has a striking effect on the ability of virus to infect them. As can be seen in figure 4, starting from immediately after heat treatment and through at least the next 9 h, there is a marked reduction in infectivity of such cells, particularly during the first 6 h.

The use of hyperthermia in conjunction with chemotherapeutic drugs has been and continues to be an area of interest for laboratory and clinical studies aimed at improving cancer therapy [34]. There is evidence from in vitro studies in our laboratory that such combined treatments are more effective than heat alone. The effect of these two agents is at least additive. This finding is not unexpected as previous studies indicate that heat prevents retroviral infection by inactivating reverse transcriptase [35, 36].

On the surface, our observation might appear trivial, as higher doses of zidovudine might accomplish the same results, and it may not be feasible to combine heat and

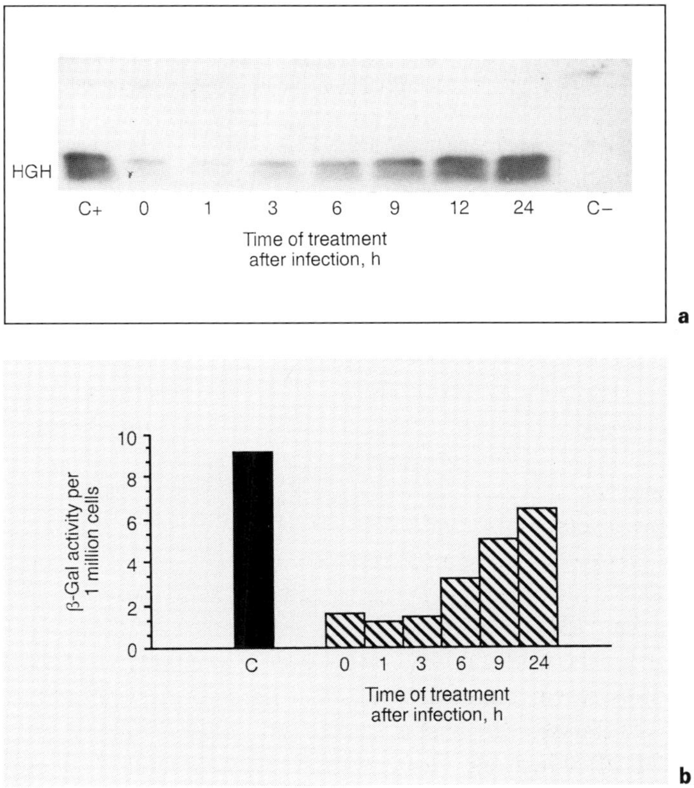

Fig. 3. a Infection before 42.8 °C for 135 min. NIH/3T3 cells were exposed to virus containing the human growth hormone (HGH) gene for 2 h and rinsed. At various times, starting immediately after infection, the cells were heated (42.8 °C/135 min). Infection was monitored in all cells 36 h after rinsing. The HGH content of 5 × 10^5 cells is depicted on Western blots using an antibody to HGH. **b** Viral infection followed by 42.8 °C for 135 min. The same experimental conditions were followed except that the virus had β-galactosidase (β-gal) integrated as its reporter gene. β-gal activities were determined by a colorimetric assay. In every case heating of the cells was staggered so all treatment groups were simultaneously exposed to the same virus preperation.

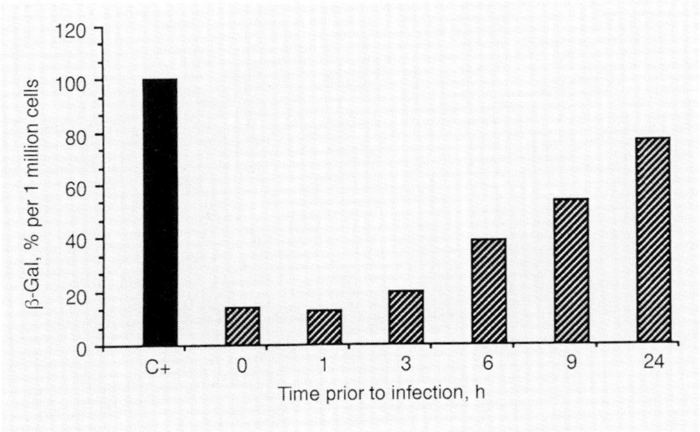

Fig. 4. NIH/3T3 cells were heated to 42.8 °C for 135 min and returned to 37 °C. At various times from 0 to 24 h after heating the cells were exposed to virus containing a β-galactosidase reporter gene. After 2 h the virus was rinsed off the cells, and after 36 h the cells were assayed for β-galactosidase. Exposure to the virus preparation is as described in figure 2.

Table 1. Effect of heating (42.8 °C) time on virus production

Heating time, min	Production[1]
0	1.00
25	1.83
75	0.97
135	0.16

[1] From Western blots of a 24-hour virus collection [Herman and Yatvin, unpublished results].

drug over an extended period of time. It is logical to ask what is the relevance of such combinations to treatment of HIV infection? The answer, we feel, is that it is a useful approach in cases of known exposure, such as accidental needle sticks. As noted by Fazely et al. [37], the rising number of people infected with HIV and their increased need for medical care greatly increases the probability of health care workers being exposed to infectious body fluids. In response to this situation, some institutions have initiated prophylactic zidovudine programs for accidentally exposed individuals. Although the number of cases is small, zidovudine alone proved ineffective in preventing infection in 3 individuals known to have been exposed to contaminated blood [38–40].

There have been numerous animal studies to determine the efficacy of prophylactic zidovudine both alone and in combination with interferon-α [41–43]. Although these animal studies have met with varying degrees of success, their relevance to the human situation remains questionable, as only cell-free virus was injected. In a real clinical situation, exposure would include infected cells. A study to determine the prophylactic effects, if any, of hyperthermia and drugs in animals, preferably primates, exposed to both free virus and infected cells is needed.

Heat and Virus Production

HIV-infected human U1 promonocytic cells exposed to heat shock for 2 h at 41.7 °C, or human ACH-2 T-lymphocytes for 80 min at 42.6 °C, upregulate virus production [3]. Others have previously reported similar observations using different virus-infected cell lines. Replication of human cytomegalovirus (CMV) was enhanced in human embryo fibroblasts heated at 44 °C for 10 min [44]. The early transcription unit of human CMV in a rat cell line was activated by exposing the cell to 42 °C for 15 min [45] and the HIV long terminal repeat (LTR) was induced by heat shock [46].

The ability of heat to modify virus production varies with both the temperature and time that infected cells are heated, and would account for the reports indicating that heat can inhibit virus production. For example, heating cultured cells at 42 °C for 1–2 h reduced mouse mammary tumor virus RNA content to almost nondetectable levels [47].

As seen in table 1, we found that heating cells producing Moloney murine leukemia virus for 135 min at 42.8 °C markedly inhibited virus production, while 75 min at the same temperature has no effect. However, after a 25-min exposure there was a marked upregulation of virus production. The virions produced at each of these temperatures were equally capable of infecting 3T3 cells.

We most emphatically do not agree with the panel and Stanley et al. [3] who imply that upregulation of virus expression by heat is, by definition, bad. Enhanced release of virus by heat is not necessarily detrimental, as there are only a finite number of cells capable of being infected. Even in patients where viremia exists, there are relatively very few infected cells [22, 48]. Thus, even if viremia were to occur as a result of hyperthermia, it would be of limited significance. Furthermore, no consideration was given to the possibility that upregulated cells may be more heat-sensitive [7–9]. Finally, in the unlikely event that viremia does turn out to be a problem, circulating virus could be removed by passing a patient's blood

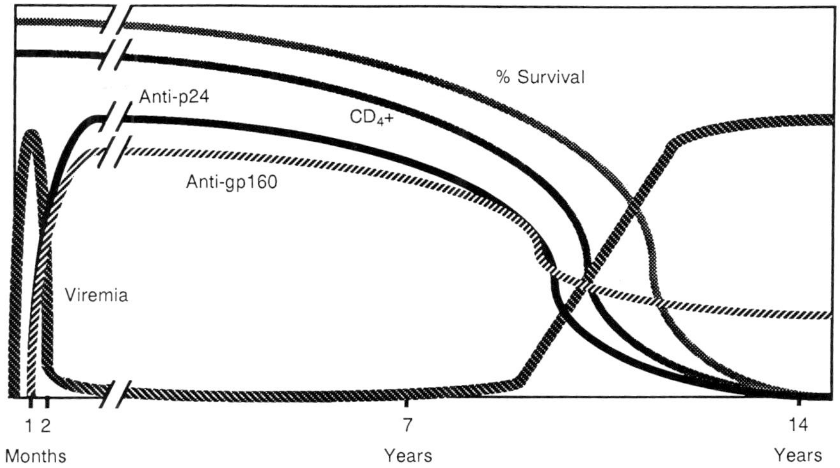

Fig. 5. Progression of HIV infection, created from the data of Clark et al. [49] and Daar et al. [50]. It shows many of the salient features of the progression of disease from infection to AIDS. It does not, however, adequately address the very early and marked reduction in CD4+ T cells, memory T cells and T-cell reactivity which never completely recover [51]. Refer to the text for a discussion of these changes reported by Miedema et al. [51].

through an affinity column prepared with antibodies to HIV proteins to remove both free virus and cells expressing viral proteins. The timing of any such treatment would be dependent on when any viremia occurred. In the study by Stanley et al. [3] there was a lag in upregulation, and virus production did not peak until day 4 or later after heating. Ideally, if any removal of circulating virus is required, it could be correlated with subsequent hyperthermia treatments.

Up to this point we have provided evidence that heat is preferentially cytocidal to HIV-infected cells, is additive with zidovudine and chaotropic agents such as ethanol, and can inactivate free virus. The potential role of heat and other agents such as lymphokines in unmasking immunologically silent cells for destruction will be covered in subsequent sections.

Heat and the Immune System

Upregulation of virus production contributes significantly to our argument about why the panel was incorrect when they stated that there is no immunological support for hyperthermia in the treatment of HIV infections. In our opinion this conclusion by the panel represents either a misreading or ignorance of information available at the time. The available evidence suggests that hyperthermia has a significant role to play as an enhancer of immunologic responses at various stages in the course of HIV infection. The evidence that heat could be an important factor in 'unmasking' immunologically silent cells for destruction is discussed below.

For the purposes of this discussion it is important for the reader to recall the natural history of HIV infection. As can be seen from figure 5, which is derived primarily from the reports by Clark et al. [49] and Daar et al. [50], primary HIV infection frequently follows an acute, self-limiting syndrome, which is characterized by high levels of infectious HIV-1 in plasma and peripheral-blood mononuclear cells (PBMC). As early as 10 days after infection, the HIV load in both plasma and PBMC spontaneously and precipitously falls two orders of magnitude to undetectable levels. At approximately the same time as the antigen and viremia levels fall, the patients seroconvert to HIV+.

Both Clark et al. [49] and Daar et al. [50] respectively draw attention to the fact that 'the rapid and spontaneous decline in the viral load suggests an effective immune response in the host that, if understood, may be used to combat AIDS', and 'Effective control of HIV-1 replication during primary infection implies the activation of clinically important mechanisms of immune defense that merit further examination in relation to ... anti-viral therapy ...'. Miedema et al. [51] have reported that CD4+ T cells, memory T cells and T-cell reactivity, which drop precipitously during the primary phase of infection, do not completely recover to their pre-infection level. In the case of CD4+ T cells the recovery reaches 75% of the pre-infection level and only about 50% for memory T cells and T-cell reactivity.

These workers describe the pathogenesis of AIDS in terms of a dynamic interaction between HIV and the immune system. Their model for AIDS pathogenesis is as follows: the asymptomatic state is initially maintained by immune suppression of HIV; persistent low-grade HIV infection increases CD4+ cell turnover with a selective loss of memory T cells; a critical loss of immune reactivity is eventually reached and the emergence and overt replication of virulent HIV variants which induce progressive depletion of CD4+ T cells and the development of AIDS. Miedema et al. [51], Clark et al. [49] and Daar et al. [50] all suggest that manipulation of the immune response in HIV-infected people could be used as a therapeutic approach to combat the onset of AIDS.

We concur, and believe that it is during the intervening period of months to years following primary HIV infection and prior to the onset of ARC and AIDS that hyperthermia has the greatest role to play. Although immune status is less than optimal, hyperthermia could enhance its effectiveness and would be helped, at least in part, by its ability to induce heat stress proteins (Hsp) and virus proteins in 'quiescent' cells. Although it is not clear why so many HIV-infected cells do not produce virus, similar findings have been reported in murine cells genetically engineered using retroviral vectors. It may be because retroviral regulatory elements are repressed in certain cell types [52, 53]. Stanley et al. [3] provide evidence that heat-induced upregulation of virus production is mediated by action on the viral LTR, which in turn is most likely mediated by the NF-kB-like enhancer sequence [3].

The expression of HIV in lymphokine-activated T cells, which increases up to 50-fold, is correlated with induction of a DNA-binding protein indistinguishable from NF-kB, which probably activates dormant HIV-infected T cells by binding to the viral enhancer region [54]. Others have reported similar lymphokine-induced amplified levels of viral replication in T cells [55]. Virus can be induced in latently HIV-infected T cells and primary mononuclear phagocytes [56] either by lymphokines or compounds that activate the cells to secrete them [57, 58]. The presence of inducible NF-kB-specific factors in monocytes and macrophages may also contribute to HIV replication [55]. Heating dormant HIV-infected cells also induces virus production via NF-kB pathways [3].

Concern has been expressed that NF-kB stimulation might pose a risk by activating dormant infected T cells and accelerating virus spread in the early stages of HIV disease [54]. Although increased risk is possible, it is unlikely since, even in the presence of massive viremia, very few T cells are infected [22, 48]. We believe that the current approach of trying to spare CD4+ and other HIV-infected cells in individuals early during the asymtomatic stage of HIV disease is counter-productive. The strategy of trying to kill diseased cells during the early stages of disease is preferable in that it presents the possibility of eliminating an HIV infection. As noted above, we believe that patients with relatively intact immune systems will be able to eliminate such unmasked cells. That belief is strongly supported by the information provided above and following.

There is increasing interest in and recognition of the fact that heat stress proteins are antigenic and play an important role in infection and immune surveillance. Young and Elliot [59] argue that individuals have the capacity to respond to self stress protein determinants in a way that would help then eliminate their own infected cells. For example, Garry et al. [60] have shown that induction of these proteins occurs in eukaryotic cells following in vitro virus infection, and Jindal and Young [61] have demonstrated that infection of human monocyte-macrophages causes a substantial increase in Hsp70 expression. Recognition of neoantigen on these cells by cytotoxic T lymphocytes (CTLs) and their elimination could delay virus spread. Koga et al. [62] have shown that primary murine macrophages infected with CMV were susceptible to killing by class-1 MHC-restricted, CD8+ CTLs specific for epitopes of *M. tuberculosis* Hsp60. Recently it has been demonstrated that deoxyspergualin, a potent immunosuppressant, specifically interacts with a member of the Hsp70 family [63]. The immune response to postinfection expression of high levels of Hsp and subsequent cell surface recognition by CTLs may limit viral spread and could be the reason the primary stage of infection is self-limiting.

The obvious question is how would WBH prove useful in treating HIV infection once viremia and high levels of viral antigen have disappeared, CD4+ cell levels have rebounded and seroconversion has occurred. The answer lies in the fact that, during the long symptom-free period following initial infection and prior to ARC or AIDS, relatively few infected cells are actively producing virus. Even during the late stages of infection less than 1 in 10 infected cells is producing virus [48]. Hence, in unheated cells the ability of MHC class-1-restricted CTLs to recognize virus-infected cells via Hsp markers is limited. Thus the use of WBH early, prior to ARC or AIDS could unmask 'quiescent' cells. In most retroviral infections virus-specific CTL surveillance keeps the infection in check [64]. Heat induction of viral proteins and Hsp would overcome the ability of these quiescent HIV-infected cells to escape CTL detection. Such treatment is more likely to be effective early in HIV infection because it is known that the virus-specific responses of CTLs are lost in patients with advanced HIV infections [65–67]. In addition, there would be added benefits from WBH as a result of their markedly altered plasma membranes [19–22] hyperthermia would more readily kill upregulated cells.

Immunomodulators have been used previously in both cancer therapy and organ transplantation and have been investigated for their potential usefulness against retroviral diseases [68, 69]. Using the mouse model system, MAIDS, Funk et al. [69] found that the drug IMEXON reversed some but not all of the symptoms. IMEXON did prevent hypergammaglobulinemia and splenomegaly and prolonged survival; however, it did not reduce serum viral titers [69]. Nevertheless, the use of heat, in combination with immunomodulators, may be worth considering.

Possible Dangers of WBH Treatment

In contrast to the assertions of Dr. Fauci, the safety of extracorporeal WBH is well established. Parks et al. [70] exposed 25 patients with preterminal bronchogenic carcinoma to extracorporeally induced WBH (41.5–42.0 °C). These very debilitated patients were given a total of 97 treatments which averaged 5 h in duration. From this series three treatment-related deaths occurred. Eisler et al. [71] treated 28 patients with 41.8 °C and encountered few complications, suggesting that WBH could be employed even in patients considered at high risk for general anesthesia. In another study by Eisler [personal commun.] 150 WBH procedures were performed on cancer patients. Some patients were heated for up to 10 h at 41.8 °C, and others for 2 h at 42.5 °C. From this series of WBH treatments 3 deaths were treatment related. They occurred either during or shortly after treatment and were attributable in two instances to toxic reactions as a result of a heat-induced rapid dissolution of the patient's tumor, and one instance of heart failure.

Conclusions and Suggestions for Using WBH to Treat HIV Infections

On the basis of the foregoing, it is our contention that the observations of Zerbini et al. [44], Geelen et al. [45, 46] and Stanley et al. [3], regarding the role of heat in stress protein induction and the upregulation and control of virus production in HIV-infected cells, must now be viewed from a completely different perspective than they were by the NIH panel and Fauci [48]. In fact, instead of heat being a negative factor in the treatment of HIV infection, precisely the opposite should prove to be the case. The greatest benefit to the patient would be derived during the early stages of infection when their immune systems are capable of responding to heat-induced stress proteins and MHC class-1 viral epitopes. Unfortunately, much less benefit is likely to be derived by WBH treatment of severely immunosuppressed patients.

We have provided the following information which supports the idea that WBH would have a beneficial action against retroviral infection. (1) Acutely and chronically virally infected cells have abnormal Golgi and plasma membranes with increased permeability which are likely to make them more heat sensitive. (2) Golgi in virally infected cells are more disrupted by heat than those in uninfected cells. (3) Heat modifies viral protein processing. (4) Heat selectively kills virally infected cells. (5) Heat kills free virus. (6) Heat effects on cells and viruses are enhanced by chaotropic (disordering) membrane-active agents. (7) Heat has prophylactic actions against acutely infected cells and is at least additive with zidovudine. (8) Heat modifies subsequent infectability of cells exposed to hyperthermia. (9) Heat enhances immune surveillance by activating quiescent infected cells that are otherwise immunologically 'invisible'. (10) WBH is well tolerated even by debilitated patients.

References

1 Yatvin MB: An approach to AIDS therapy using hyperthermia and membrane modification (abstract). 4th Annu Meet Am Soc Clin Hyperthermic Oncol, Houston, Nov 1987.
2 Yatvin MB: An approach to AIDS therapy using hyperthermia and membrane modification. Med Hypoth 1988;27:163–165.
3 Stanley SK, Bressler PB, Poli G, Fauci AS: Heat shock induction of HIV production from chronically infected promonocytic and T cell lines. J Immunol 1990;145:1120–1126.
4 Hilts PJ: Panel finds heating blood is useless in treating AIDS. New York Times, Sept 5, 1990, p A16.
5 Robins HI, Dennis WH, Martin PA, Sondel PM, Yatvin MB, Steeves RA: Potentiation of differential hyperthermic sensitivity of AKR leukemia and normal bone marrow cells by lidocaine or thiopental. Cancer 1984;54:2831–2835.
6 Robins HI, Sielaff KM, Storer B, Hawkins MJ, Borden EC: Phase I trial of human lymphoblastoid interferon with whole body hyperthermia in advanced cancer. Cancer Res 1989;49:1609–1615.
7 Wong GHW, McHugh T, Goeddel DV: HIV-infected cells are more susceptible than uninfected cells to killing by irradiation, heat or tumor necrosis factor. 2nd Int Conf on Drug Res in Immunology and Infectious Disease: Aquired Immunol deficiency Syndrome (AIDS) Nov 1989.

8 Omar RA, Yano S, Kikkawa Y: Antioxidant enzymes and survival of normal and simian virus 40-transformed mouse embryo cells after hyperthermia. Cancer Res 1987;47:3473–3476.
9 Wong GHW, McHugh T, Weber R, Goeddel DV: Tumor necrosis factor a selectively sensitizes human immunodeficiency virus-infected cells to heat and radiation. Proc Natl Acad Sci USA 1991;88:4372–4376.
10 Shaw GM, Hagn BH, Arya SK, Groopman JE, Gallo RC, Wong-Staal F: Molecular charaterization of human T-cell leukemia (lymphotropic) virus type III in the acquired immune deficiency syndrome. Science 1984;226:1165–1171.
11 Keshet E, Temin HM: Cell killing by spleen necrosis virus is correlated with a transient accumulation of spleen necrosis virus DNA. J Virol 1979;31:376–388.
12 Mullins JI, Chen CS, Hover EA: Disease-specific and tissue-specific production of unintegrated feline leukaemia virus variant DNA in feline AIDS. Nature 1986;319:333–336.
13 Koga Y, Lindstrom E, Fenyo EM, Wigzell H, Mak TW: High levels of heterodisperse RNAs accumulate in T cells infected with human immunodeficiency virus and in normal thymocytes. Proc Natl Acad Sci USA 1988;85:4521–4525.
14 Somasundaran M, Robinson HL: Unexpectedly high levels of HIV-1 RNA and protein synthesis in a cytocidal infection. Science 1988; 241:1554–1557.
15 Clark AW, Robins HI, Vorphal JW, Yatvin MB: Structural changes in murine cancer associated with hyperthermia and lidocaine. Cancer Res 1983;43:1716–1723.
16 Duskin D, Holbrook K, Williams K, Bornstein P: Cell surface morphology and adhesive properties of normal and virally transformed cells treated with trincamycin, an inhibitor of protein glycosylation. Exp Cell Res 1978;116:153–165.
17 Whaley WG, Leffingwell TP, Dauwalder M: Oncornavirus modification of the Golgi apparatus. Cytobios 1979;22:105–131.
18 Bestwick RK, Kozak SL, Kabat D: Overcoming interference to retroviral superinfection results in amplified expression and transmission of cloned genes. Proc Natl Acad Sci USA 1988; 85:5404–5408.
19 Yamaizumi M, Uchida T, Odada Y: Macromolecules can penetrate the host cell membrane during the early period of incubation with HVJ (Sendai virus). Virology 1979;95:218–221.
20 Fernandez-Puentes C, Carrasco L: Viral infection permeabilizes mammalian cells to protein toxins. Cell 1980;20:769–775.
21 McGrath MS, Hwang KM, Caldwell SE, Gaston I, Luk K-C, Wu P, Ng VL, Crowe S, Daniels J, Marsh J, Deinhart T, Lekas PV, Vennari JC, Yeung H-W, Lifson JD: GLQ223: An inhibitor of human immunodeficiency virus replication in acutely and chronically infected cells of lymphocyte and mononuclear phagocyte lineage. Proc Natl Acad Sci USA 1989;86:2844–2848.
22 Zarling JM, Moran PA, Haffar O, Sias J, Richman DD, Spina CA, Myers DE, Kuebelbeck V, Ledbetter JA, Uckum FM: Inhibition of HIV replication by pokeweed antiviral protein targeted $CD4^+$ cells by monoclonal antibodies. Nature 1990;347:92–95.
23 Leonard R, Zagury D, Desportes I, Bernard J, Zagury J-F, Gallo RC: Cytopathic effect of human immunodeficiency virus in T4 cells is linked to the last stage of virus infection. Proc Natl Acad Sci USA 1988;85:3570–3574.
24 Lynn WS, Tweedale A, Cloyd MW: Human immunodeficiency virus (HIV-1) cytotoxicity: Perturbation of the cell membrane and depression of phospholipid synthesis. Virology 1988; 163:43–51.
25 Yatvin MB, Dennis WH, Elegbede JA, Elson CE: Sensitivity of tumor cells to heat and ways of modifying the response; in Bowler K, Fuller BJ (eds): Temperature and Animal Cells. Cambridge, Company of Biologists, 1987, pp 235–267.
26 Yatvin MB, Cramp WA: The role of cellular membranes in hyperthermia: some observations and theories reviewed. Int J Hyperthermia 1993;9:165–185.
27 Gordon LM, Jensen FC, Curtain CC, Mobley PM, Aloia RC: Thermotropic lipid phase separation in the human immunodeficiency virus. Biochim Biophys Acta 1988;943:331–342.
28 McDougal JS, Martin LS, Cort SP, Mozen M, Heldebrant CM, Evatt BL: Thermal inactivation of the acquired immunodeficiency syndrome virus human T lymphotropic virus-III/lymphadenopathy-associated virus, with special reference to antihemophilic factor. J Clin Invest 1985;76:875–877.
29 Waite MRF, Pfefferkorn ER: Inhibition of Sindbis virus production by media of low ionic strength: intracellular events and requirements for reversal. J Virol 1970;5:60–71.
30 Wallis C, Melnick JL, Rapp F: Different effects of $MgCl_2$ and $MgSO_4$ on the thermostability of viruses. Virology 1965;26:694–699.
31 Snipes W, Person S, Keith A, Cupp J: Butylated hydroxytoluene inactivates lipid-containing viruses. Science 1975;188:64–65.
32 Moore MF, Patzer EJ, Shaw JM, Thompson TE, Wagner RR: Interaction of vesicular Sto virus with lipid vesicles: Depletion of cholesterol and effect on virion membrane fluidity and infectivity. J Virol 1978;27:320–329.
33 Aloia RC, Jensen FC, Curtain CC, Mobley PW, Gordon LM: Lipid composition and fluidity on the human immunodeficiency virus. Proc Natl Acad Sci USA 1988;85:900–904.
34 Hahn GM: Potential for therapy of drugs and hyperthermia. Cancer Res 1979;39:2264–2268.
35 Bassin RH, Duran-Troise G, Gerwin BI, Rein A: Abrogation of $Fv\text{-}1^b$ restriction with murine leukemia viruses inactivated by heat or by gamma irradiation. J Virol 1978;26:306–315.
36 Hirai R, Yuasa Y, Yamamoto T: Heat-labile character of murine sarcoma-xenotropic leukemia virus complex in duck cells. Virology 1979; 96:615–621.
37 Fazely F, Haseltine WA, Rodger RF, Ruprecht RM: Postexposure chemoprophylaxis with ZDV or ZDV combined with interferon-α: Failure after inoculation Rhesus monkeys with a high dose of SIV. J Acquir Immune Defic Syndr 1991;4:1093–1097.
38 Looke DEF, Grove DI: Failed prophylactic zidovudine after needlestick injury. Lancet 1990;335:1280.
39 Lange JMA, Boucher CAB, Hollak CEM: Failure of zidovudine prophylaxis after accidental exposure to HIV-1. N Engl J Med 1990;322: 1375–1377.
40 Durand E, Le Jeune C, Hughes F-C: Failure of prophylactic zidovudine after suicidal self-inoculation of HIV-infected blood. N Engl J Med 1991;324:1062.
41 Ruprecht RM, Chou T-C, Chipty F: Interferon-α and 3'-azido-3'-deoxythymidine are highly synergistic in mice and prevent viremia after acute retrovirus exposure. J Acquir Immune Defic Syndr 1990;163:581–600.
42 Shih C-C, Kaneshima H, Rabin L: Postexposure prophylaxis with zidovudine suppresses human immunodeficiency virus type 1 infection in SCID-hu mice in a time-dependent manner. J Infect Dis 1991;163:625–627.
43 Kaneshima H, Shih C-C, Namikawa R: Human immunodeficiency virus infection of human lymph nodes in the SCID-hu mouse. Proc Natl Acad Sci USA 1991;88:4523–4527.
44 Zerbini M, Musiani M, La Placa M: Stimulating effect of heat shock on the early stage of human cytomegalovirus replication cycle. Virus Res 1986;6:211–216.
45 Geelen JLMC, Boom R, Klaver GPM, Minnaar RP, Festkamp MCW, Van Milligen FJ, Sol CJA, Van Der Noordaa J: Transcriptional activation of the major immediate early transcription unit of human cytomegalovirus by heat-shock, arsenite and protein synthesis inhibitors. J Gen Virol 1987;68:2925–2931.
46 Geelen JLMC, Minnaar RP, Boom R, Van Der Noordaa J, Goudsmit J: Heat-shock induction of the human immunodeficiency virus long terminal repeat. J Gen Virol 1988;69:2913–2917.
47 Ralhan RJ, Johnson GS: Destabilization of cytoplasmic mouse mammary tumor RNA by heat shock: prevention by cycloheximide pretreatment. Biochem Biophys Res Commun 1986;137:1028–1033.
48 Fauci AS: Pharmacologic regulation of HIV expression. AIDS Res Human Retroviruses 1992;8:707–710.
49 Clark SJ, Saag MS, Don Decker WD, Campbell-Hill S, Roberson JL, Veldkamp PJ, Kappes JC: High titers of cytopathic virus in plasma of patients with symptomatic primary HIV-1 infection. N Engl J Med 1991;324:654–960.
50 Daar ES, Moudgil T, Meyer RD, Ho DD: Transient high levels of viremia in patients with primary human immunodeficiency virus type 1 infection. N Engl J Med 1991;324:961–964.

51 Miedema F, Tersmette M, van Lier RAW: AIDS pathogenesis: a dynamic interaction between HIV and the immune system; in Gallagher RB (ed): HIV and the Immune System. Amsterdam, Elsevier, 1991, pp 89–97.
52 Hoeben RC, Migchielsen AAJ, Van Der Jagt RCM, Van Ormondt H, Van Der EB AJ: Inactivation of the moloney murine leukemia virus long terminal repeat in murine fibroblast cell lines is associated with methylation and dependent on its chromosomal position. J Virol 1991;65:904–912.
53 Xu L, Yee J-K, Wolff JA, Friedmann T: Factors affecting long-term stability of moloney murine leukemia virus-based vectors. Virology 1989;171:331–341.
54 Nabel G, Baltimore D: An inducible transcription factor activates expression of human immunodeficiency virus in T cells. Nature 1987; 326:711–713.
55 Greene WC, Bohnlein E, Ballard DW: HIV-1, HTLV-1 and normal T-cell growth: transcriptional strategies and surprises. Immunol Today 1989;10:272–278.
56 Koyanagi Y, O'Brien WA, Zhao JQ, Golde DW, Gasson JC, Chen ISY: Cytokines alter production of HIV-1 from primary mononuclear phagocytes. Science 1988;241:1673–1675.
57 Zagury D, Bernard J, Leonard R, Cheynier R, Feldman M, Sarin PS, Gallo RC: Long-term cultures of HTLV-III-infected T cells: A model of cytopathology of T-cell depletion in AIDS. Science 1986;231:850–853.
58 Harada S, Koyanagi Y, Nakashima H, Kobayashi N, Yamamoto N: Tumor promoter, TPA, enhances replication of HTL-III/LAV. Virology 1986;154:249–258.
59 Young RA, Elliott TJ: Stress proteins, infection, and immune surveillance. Cell 1989;59: 5–8.
60 Garry RF, Ulug ET, Rose HR: Induction of stress proteins in sindbis virus- and vesicular stomatitis virus-infected cells. Virology 1983; 129:319–332.
61 Jindal S, Young, RA: Vaccinia virus infection induces a stress response that leads to association of Hsp with viral proteins. J Virol 1992;66: 5357–5362.
62 Koga T, Wand-Wurttenberger A, Debruyn J, Mund ME, Schoel B, Kaufmann SHE: T cells against a bacterial heat shock protein recognize stressed macrophages. Science 1989;245:1112–1115.
63 Nadler SG, Tepper MA, Schacter B, Mazzucco CE: Interaction of the immunosuppressant deoxyspergualin with a member of the hsp70 family of heat shock proteins. Science 1992; 258:484–486.
64 Chieco-Bianchi L, Collavo D, Biasi G: Immunologic unresponsiveness to murine leukemia virus antigens mechanisms and role in tumor development. Adv Cancer Res 1988;51:277–306.
65 Shearer GM, Bernstein DC, Tung KSK, Via CS, Redfield R, Salahuddin SZ, Gallo RC: A model for the selective loss of major histocompatibility complex self-restricted T cell immune responses during the development of acquired immune deficiency syndrome (AIDS). J Immunol 1986;137:2514–2521.
66 Clerici M, Stocks NI, Zajac RA, Boswell RN, Lucey DR, Via CS, Shearer GM: Detection of three distinct patterns of T helper cell dysfunction in asymptomatic, human immunodeficiency virus-seropositive patients. J Clin Invest 1989;84:1892–1899.
67 Pantaleo G, De Maria A, Koenig S, Butini L, Moss B, Baseler M, Lane HC, Fauci AS: $CD8^+$ T lymphocytes of patients with AIDS maintain normal broad cytolytic function despite the loss of human immunodeficiency virus-specific cytotoxicity. Proc Natl Acad Sci USA 1990;87: 4818–4822.
68 Sizemore R, Deinglewicz R, Garry R, Sinha S: IMREG-1, a leukocyte-derived immunomodulator, does not enhance HIV replication in AIDS and ARC patients. Fed Am Soc Exp Biol J 1988;2:830.
69 Funk CY, Eisman J, Hersh EM: Treatment of the murine, retrovirus-induced lymphoproliferative immunodeficiency disease (LP-BM5) in C57BL/10 mice with the immunomodulator IMEXON. AIDS Res Hum Retroviruses 1992; 8:633–638.
70 Parks L, Minaberry D, Smith D, Neely W: Treatment of far-advanced bronchogenic carcinoma by extracorporeally induced systemic hyperthermia. J Thorac Cardiovasc Surg 1979; 78:883–892.
71 Eisler K, Landauer B, Hipp R, Kolb E, Lange J, Siewert JR, Zanker K, Blumel G: Experiences with therapeutic whole-body hyperthermia. Anaesthesist 1985;34:299–303.

Oncology 1993;50:390–392

G.M. Pigliucci
F. Giudice Andrea
D. Venditti
V. Cervelli
C.U. Casciani

Department of Surgery, II University of Rome 'Tor Vergata', Rome, Italy

Optimization of Pre-, Intra- and Postoperative Hyperthermic Treatment in Inoperable Lower Bowel and Liver Tumors

Key Words
Hyperthermia
Liver tumors
Lower bowel tumors

Abstract
Traditional therapies in malignant tumors are often inadequate, either because the tumors aren not responsive or because the tolerance dosage has little effect. With this in mind, our study examined advanced tumors of the lower bowel and liver. We evaluated the results of 10 cases of anorectal epidermoid cancer. We treated these patients with radiofrequency hyperthermia, since previous radiotherapy was unsuccessful. The patients had significant tumor remission, along with a 'free-of-disease' follow-up of 10 months. In this paper, we also report the positive response in 10 cases of multifocal liver tumors. All these tumors had metastasized from gastrointestinal sites and were not responsive to traditional therapies. These patients had remission with a follow-up of 6 months. In the first cases, we associated 'low-dose' radiotherapy with hyperthermia; in the latter cases, we induced chemoembolization and also associated it with hyperthermia. We emphasize these associations and suggest the very important role of intraoperative hyperthermia in absolutely inoperable cases or in those cases previously diagnosed as unresectable.

Today it is possible to observe how traditional therapies of malignant tumors often present rather insurmountable limits consisting of: tumor diffusion (local or general); systemic conditions of inoperability, or intolerance to chemotherapeutic drugs and radiotherapy.

On the other hand, the concept that has become more accepted in modern oncology defines the tumor as a disease affecting the whole organism and its immunological mechanisms, and not solely as an organic disease. This is especially valid for certain tumors which have a tendency to spread and to metastasize early (lymphomas, microcytomas, tumors of the breast, etc.) In the endeavor to overcome this 'therapeutic stasis', new complimentary methods aim at minimizing the immunodepressive factors (high dosages of polychemotherapy, radiotherapy sometimes employed without specific indications), and at eventually associating methods or substances at reduced dosages in order to synergetically potentiate the positive outcome of traditional techniques.

While it is a fact that in the last few years the concept of radical surgery has been redefined, it is equally true that surgery can undergo a synergism in all its effects by a correct association with pre-, intra- or postoperative complementary therapies. Hyperthermia, specific monoclonal antibodies, interleukin-2 and interferon, these are the methods/substances upon which the surgeon is currently placing his hope and his trust in, apart from his own experience and ability, in order to obtain an increase in the

G.M. Pigliucci, MD
Department of Surgery
II University of Rome
Via Lutezia, 2
I–00198 Rome (Italy)

percentage of partial or total malignant tumor remissions and an amelioration of the quality of life.

The period of evisceration and of grossly deforming surgery, erroneously defined as radical, is over. Today, with the aid of DNA aneuploidy study (cytofluorometry), these earlier techniques have been shown to be nothing more than exercises in anatomical surgery. In fact, with DNA cytofluorometry, it has been possible to identify cells with high neoplastic potential in apparently healthy tissues and in those distant from the excised tumor.

A comment is necessary on the negative influence that prolonged general anesthesia has on the immune response; in particular, it is already known that anesthesia, especially if long-term, can depress the lymphocyte T response by altering both the production as well as the opsonization property of phagocytes.

In our opinion, the best way to attack a malignant tumor within the limits of operability, or clearly inoperable, is by cytoreduction in such a manner as to leave tumor remnants that can be more easily attacked by the associated therapies. As occurs in radiotherapy and hyperthermia, intraoperative treatment undoubtedly helps (as is already the case in institutions where it is practiced) by acting directly in or on the tumor by using apparatus that produce hyperthermia, as already occurs in interstitial hyperthermic therapy which allows one to reach effective temperatures without side effects. In this way, many of the problems that limit its use externally on deep tumors ar overcome (dyshomogeneity of tissues, different absorption by the liquid and solid components in relation to the hyperthermic heat, scarce ability to 'heat' deep tumors by means of certain apparatus for hyperthermia). With respect to the organs and systems that can benefit from the hyperthermic-surgical association, we wanted to turn our attention to tumors of the lower bowel and liver.

The following are therapeutic procedures for inoperable liver tumors which benefit from the use of hyperthermia, primarily with regard to those tumors in which it was impossible to perform radical extirpation and those diffuse tumors in which it is possible to catheterize the hepatic artery; thereby associating regional chemotherapeutic infusion or chemoembolization to the main procedure. Such associations are naturally individualized in accordance with the hepatic function of the patient and with the location of the tumor. This intra-arterial chemotherapy encounters limitations in the pre-existence of a hepatopathy, ascites and extrahepatic tumor diffusion.

The problem of measuring the temperature in hepatic tumors can be solved by inserting thermocouples in the neoplasm and maintaining them in place throughout the entire period of hyperthermia. Of particular interest in this field is the prospect, already a reality in some institutions, of eventually associating intraoperative hyperthermia with intraoperative radiotherapy. With regard to this technique, one can utilize microwave apparatus which operate at low depth, thus requiring direct contact with the tumor mass. Since intraoperative hyperthermia is a recent therapeutic method, the results thus far obtained are still subject to discussion, as are the procedures which foresee its use – but the prospects are encouraging. With regard to 'closed abdomen' hyperthermia as adjuvant or neo-adjuvant to surgery, this can be created by apparatus at a frequency of 8–13.56 MHz, utilizing a generator that produces electromagnetic waves which can bring a therapeutic temerature of 41–42 °C to a depth of 7–8 cm. Generally, hyperthermic sessions take place 3–4 times, separated one from the other by at least 48–72 h. If hyperthermic treatment is carried out in association with regional chemotherapy, and if the latter is administered by means of an intra-arterial pump, then hyperthermic treatment is contemporaneous. If, on the other hand, chemoembolization is implemented by the administration of a chemotherapeutic 'bolus', it is advisable to begin hyperthermia 4–10 days later.

According to Nagata et al. [1], hyperthermic treatment associated with regional chemoembolization brought about tumor remission in approximately 35% of the cases. According to Hamazoe et al. [2], the response to combined treatment in 11 patients was 55% (complete and partial remission) with a mean survival time of 23 weeks.

In our experience with 10 cases treated with this procedure, we had a partial response in 40% of the cases, with a mean survival time of 8 months None of our cases achieved complete remission, but we obtained stabilization of the tumor situation in 20% of the cases.

Recently employed in certain unique inoperable liver lesions is the triple association of radio- and chemotherapy and hyperthermia. The preliminary results appear even more encouraging, since there are fewer side effects present due to the lower dosages of radiation and chemotherapeutic drugs used. Regarding the pathology of the large intestine, we feel it is particularly useful to discuss inoperable anorectal 'ab initio' or relapsing tumors.

Anorectal neoplasms constitute a pathology that can become rapidly inoperable and which often present resistance to isolated chemo- and radiotherapy. On the other hand, the clinical aspects of these tumors become so dramatic and important for the general state of the patient that they often lead to a decline in his general condition (pain, hemorrhage, fistulation, infiltration of contiguous tissues, etc.). The limitation of isolated radiotherapeutic treatment for prolonged periods of time in this type of patient leads us to consider the simultaneous association of radiotherapy and hyperthermia as a desirable alternative to be utilized 'ab initio'. This association foresees hyperthermic treatment with microwaves of radiofrequency, depending upon the extent and diffusion of the tumor, to be carried out 1–1.5 h after radiation treatment. Measuring the temperature in these cases is particularly easy and this allows us to support the fundamental reasons for using this treatment.

In our study of 10 cases of anorectal epidermoidal cancer (5 nontreated inoperable cases and 5 cases no longer responsive to radiation treatment), we obtained a positive response (cr + pr) in 50% of the cases with a mean survival time of 12 months.

Generally, from the viewpoint of associating hyperthermic to colorectal surgery, it is important to emphasize how hyperthermia can currently be employed in the preoperative phase. We point out how its potential application in the operative phase seems indicative in optimizing the radicality of surgery. In fact, according to us, the intraoperative use of hyperthermia constitutes an advancement in the realization of a correct and more efficient multidoctrinal approach. The direct application in the tumor of electromagnetic waves, which are therapeutically efficacious if emitted from microwave antennas, represent the most modern method of neutralizing inoperable tumors, analogous to that which takes place in intraoperative radiotherapy. In pre-operative treatment, one whishes to reduce the neoplastic mass in order to facilitate surgery. Treatment is administered at least 20 days prior to surgery, with 3 or more sessions, depending on the dimension of the neoplasm, and preferably utilizing radiofrequency apparatus given the depth of the tissues to be treated.

This pre-operative treatment has many optimistic prospects, even though at present the quantity of data to support its efficacy is scarce. In addition, it is known that the concept of pre-operative treatment is becoming more acceptable today in conjunction with traditional therapies.

With respects to postoperative hyperthermia, this is practiced when lymph node involvement (N+ tumors) exists and is possible associated with radio- or chemotherapy, depending on the location and on the histological nature of the tumor. Horie's [3] procedures call for the triple association of radio-, thermo- and chemotherapy.

In conclusion, the prospect of associating hyperthermia with other eventual therapies, complementary to surgery, can be considered as a strategy that can render certain surgery more therapeutic, especially with respect to the long-term survival. In addition, it can be considered as an integrated, feasible therapy with ever-increasing optimization of its fundamental reasons by virtue of the possibility that intraoperative hyperthermia offers us.

References

1 Nagate Y, et al: Radiofrequency thermotherapy for malignant liver tumors. Cancer 1990;65: 1730–1736.

2 Hamazoe R, Koga S, Maeta M, Murakami A, Inoue Y, Shabana M: Intra-hepato-arterial chemotherapy combined with hyperthermic treatment: Clinical results of metastatic cancer of the liver and effects on correct (but not at all necessary) hepatic blood flow. Gan To Kagaku Ryoho 1989(suppl 8, pt 2):3032–3035.

3 Horie H: Preoperative hyperthermo-chemo-radiotherapy to prevent local recurrence of rectal cancer. Lecture 12th Int Symp Oncol Clin Hyperther ICHS, Rome 1989.

4 Leveen HH: Optimization of hyperthermia. Proc 12th Symp Clin Hyperther, Rome 1989.